THE NERVOUS SYSTEM
AND ELECTRIC CURRENTS

THE NERVOUS SYSTEM
AND ELECTRIC CURRENTS

Proceedings of the Third Annual National Conference of the Neuro-Electric
Society, held in Las Vegas, Nevada, March 23-25, 1970

Edited by
Norman L. Wulfsohn

Associate Professor of Anesthesiology
University of Texas Medical School at San Antonio
San Antonio, Texas

and

Anthony Sances, Jr.

Professor and Chairman, Biomedical Engineering
Marquette University and School of Medicine
Milwaukee, Wisconsin

🏛 PLENUM PRESS · NEW YORK-LONDON · 1970

Library of Congress Catalog Card Number 70-114562

ISBN-13: 978-1-4684-1838-5   e-ISBN-13: 978-1-4684-1836-1
DOI: 13: 10.1007/978-1-4684-1836-1

© **1970 Plenum Press, New York**

Softcover reprint of the hardcover 1st edition 1970

A Division of Plenum Publishing Corporation
227 West 17th Street, New York, N.Y. 10011

United Kingdom edition published by Plenum Press, London
A Division of Plenum Publishing Corporation, Ltd.
Donington House, 30 Norfolk Street, London, W.C.2, England

PREFACE

The brain is one of the most fascinating organs of the body, which delicately controls the thoughts and activities of Man from moment to moment. While much information has been accumulated over the years as to its anatomical and physiological functions, little research has been devoted to determine the effects of electrical fields upon this organ. Therefore, a group of interested researchers formed the Neuro-Electric Society to provide a forum for studying the effects of electrical currents upon the related nuerophysio-logical determinates. Much of this research has been directed to-wards the production of sleep or a state of anesthesia by trans-cranially applied electrical currents.

The Neuro-Electric Society is a national society which holds annual conferences, bringing together a wide variety of scientists with skills in various fields such as anesthesiology,biomathematics, biomedical engineering, neuro-anatomy, neurology, neurophysiology, neurosurgery, pharmacology, psychiatry and psychology.

Subsequent to a number of meetings of various interested groups, the first annual meeting of the Neuro-Electric Society was held in Milwaukee, Wisconsin in October 1967, and the second in San Francisco, California in February 1969. The third conference is to be held at Las Vegas, Nevada, March 23-25, 1970.

The Editors wish to thank the following sponsors: Office of Naval Research, Washington, D.C.; Medtronic, Inc., Minneapolis, Minnesota; E. R. Squibb & Sons, Inc., New York, N.Y.; and Varo Inc., Dallas, Texas. We also wish to thank the Association for the Ad-vancement of Medical Instrumentation, Bethesda, Maryland and the Institute of Electrical and Electronic Engineers (Group of Engineering in Medicine and Biology), New York, for their cooperation.

Preparation was only completed because of the excellent help of Mr. E. Gelineau and Mrs. Linda Peters.

The Editors,

N. L. Wulfsohn

March 1970

A. Sances

THE NEURO-ELECTRIC SOCIETY

Officers and Conference Organizing Committee

Anthony Sances, Jr., President, Marquette University
Alan R. Kahn, Vice President, Health Technology Corporation
David V. Reynolds, Vice President, University of Windsor
Bernard Saltzberg, Vice President, Tulane University
Jack J. Snyder, Secretary, Hoffman-La Roche
James W. Prescott, Co-Chairman, National Institute of Child Health
Max Shapiro, Co-Chairman, University of California at Los Angeles
Norman L. Wulfsohn, Chairman of Clinical Sessions, University of
        Texas Medical School at San Antonio

National Advisory Board

Edgar J. Baldes   Irving Lutsky
Joseph J. Barboriak   James A. Meyer
Robert O. Becker   Donald H. Reigel
Camine D. Clemente   Alfred W. Richardson
Gilbert B. Devey   Lawrence R. Rose
Donald F. Flickinger   Charles E. Short
Lester A. Geddes   Kenneth A. Siegesmund
Ernest O. Henschel   Charles F. Stroebel
Reginald A. Herin   Calvin C. Turbes
Raymond T. Kado   Arthur S. Wilson
Sanford J. Larson

CONTENTS

SECTION   1

Neurophysiological Measurement of Electrical Parameters

A General Theory of Electromagnetic Measurements in Living
    Organisms and the Experimental System for Its Verification   3
    J. Kryspin and M. F. Roseman

Magnetic Fields Associated with Nervous Conduction   9
    A. M. Cook and F. M. Long

Current Density Measurement -- An Attempt for Improvement   .   15
    H. J. Marsoner

A Method of Evaluating Electrode Placements Used in
    Impedance Cephalography . . . .   ⋅ 21
    D. A. Driscoll

An Investigation of Excitability Thresholds Using Two
    Closely-Spaced High Frequency Sinusoids   27
    S. Rush and J. A. Abildskov

Spatio-Temporal Potential Fields of Evoked Responses
    Recorded Transcranially -- Prelude to Stimulation   33
    D. G. Childers, J. R. Bourne, and N. W. Perry

Impedance Measurements of Tissue   39
    R. Cohn and H. S. Leader

Electrical Activation of Skeletal Muscle by Sequential
    Stimulation   45
    P. H. Peckham, J. P. Van Der Meulen, and J. B. Reswick

A Theoretical Analysis of Several Electrode Placements
    on the Scalp for Recording Averaged Evoked Potentials . .   51
    D. A. Driscoll

S E C T I O N   2

Neurophysiological Effects of Electrical Currents

Effect of Microwave Radiation on the Frog Sciatic Nerve . .   57
    J. Rothmeier

Effect of Cortically Applied DC Currents on Strychnine
    Spike Activity -- A Preliminary Report   71
    H. Fukuda, H. J. Marsoner, and F. M. Wageneder

Parkinson-like Tremor Production by Transcranial
    Electrical Stimulation   79
    R. S. Pozos and J. R. Holbrook

Electrostimulation of Hearing   85
    M. Hoshiko

Gradable Pain Production   89
    R. H. Smith and J. H. Andrew

The Use of Applied DC Fields in the Analysis of Interictal
    Epileptiform Discharges   93
    C. A. Gleason

A Pilot Study of the Effects of Functional Electrical
    Stimulation on the Recovery of Function Following
    Stroke in Man   99
    P. E. Crago, J. P. Van Der Meulen, and J. B. Reswick

Intravascular Electrical Thrombogenesis -- A Preliminary
    Report   105
    M. McMinn, C. R. Youmans, Jr., Wm. Bucholtz,
    N. Welford, and J. R. Derrick

A Clinical Assessment of the Conditional Effects of
    Electroshock   111
    V. Johnson

Effect of Ethanol on Somatosensory Evoked Potentials   117
    S. A. E. Rosenthal, D. E. Dallmann, F. P. Goldstein,
    A. Sances, Jr., and S. J. Larson

A Comparative Study of Evoked Unit and Population Retinal
    Potentials During the Application of Diffuse Electrical
    Currents   125
    E. J. Zuperku, A. Sances, Jr., and S. J. Larson

SECTION   3

Electrosleep and Electro-Anesthesia

The Basic Rest-Activity Cycle (Abstract Paper)   133
    N. Kleitman

Basal Forebrain Structures and the Electrical Induction
    of Sleep (Abstract Paper)   135
    C. D. Clemente

A Preliminary Study of the Use of Electrosleep Therapy
    in Clinical Psychiatry   137
    R. R. Koegler, S. M. Hicks, L. Rogers, and J. H. Barger

Electrode·Position in Electro-Sleep   145
    N. L. Wulfsohn and L. Waldron

Electrosleep in Man by a Combination of Magneto-Inductive
    and Transtemporal Electric Currents   153
    D. P. Photiades, R. J. Riggs, S. C. Ayivorh, and
    J. O. Reynolds

Electrostimulation of Deep Brain Structures   159
    N. L. Wulfsohn, A. Davis, and E. Gelineau

Clinical Experiences: Electro-Anesthesia   165
    K. R. Rama Rao, S. Bhat, and M. S. Rajagopalan

Electroanesthesia Studies: Site of Current Action   173
    J. Tatsuno, R. H. Smith, H. Suzuki, and J. H. Andrew

Index   177

SECTION 1

NEUROPHYSIOLOGICAL MEASUREMENT OF ELECTRICAL PARAMETERS

A GENERAL THEORY OF ELECTROMAGNETIC MEASUREMENTS IN LIVING ORGANISMS AND THE EXPERIMENTAL SYSTEM FOR ITS VERIFICATION

Jan Kryspin and Morton F. Roseman

Institute of Bio-Medical Electronics

University of Toronto, Toronto, Canada

Abstract. The limitations for a straightforward application of the electromagnetic field equations to biological systems are mainly the inhomogeneity, time-variability and a multi-level type of organization. For these reasons, the quantities of the electromagnetic field do not represent a consistent system of biological properties or phenomena. At present this does not seem to be the obstacle for numerous applications of electromagnetic measurements to living organisms because the data represent mainly the relations between some physiological process and an operationally defined physical quantity. On the other hand, the mathematical models of such applications have necessarily to be of a very limited character, they have to be extensive without any possibility for a biologically meaningful minimalization and without a prospective synthetic generalization. After having developed a generalized geometrical approach to these problems we are proposing a theory that assumes the applicability of the RIEMANN's tensor to biological problems. The fundamental difference from the usual application in physics is the use of a probabilistic coordinate system that is limited to four coordinate axes and six coordinate planes; these geometrical concepts represent relations of different complexity. The "informational operator" is defined as a sequence of mathematical operations that connect the data obtained by physical measurements with the abstract representation.

Introduction. In the electromagnetic measurements on living organisms, we define operationally one primary quantity, the current, and one secondary quantity, the voltage. We cannot refer meaningfully to any other quantity (charge, resistivity, permittivity). But even the values of voltage and current have different meaning on different levels of biological organization (e.g.

membrane ionic currents on cellular level, pulse-volume current
fluctuations on the organ level). These limitations, however, do
not prevent very wide applications of electromagnetic measurements
to life phenomena, because the biologist is more interested in
their relational than "absolute" physical aspects. To integrate
the electromagnetic field aspects of measurements with their relational biological meaning we have developed the following theoretical and experimental approach.

  Theory. The geometry of a four-dimensional curved space as
represented by a contracted RIEMANN's tensor is taken as an abstract model of the totality of phenomena in nature, including the
life phenomena. The contracted RIEMANN tensor is identified with
the matter-energy tensor T_i^j (RAINICH 1950). The tensor

$$T_i^j = M_i^j - E_i^j$$

represents the generalization of the physical properties of matter
that are fundamental for all phenomena, living or non-living.
M_i^j is the tensor representing equations of continuity and motion
and E_i^j represents the generalization of MAXWELL's equations for
space with matter. A modified matter-energy tensor that accounts
for life phenomena as well is designated $*T_i^j$ and is assumed to
characterize the total field. Because it can be identified with
the contracted RIEMANN tensor
$R_i^j - 1/2\delta_{ij}R$, (where $\delta_{ij} = g^{i\alpha}g_{\alpha j}$ represents the metric of the
total field) it has only 10 physical components which are usually
recognized as density, momenta and electromagnetic quantities in
the matter-energy tensor. For the characterization of life phenomena by the tensor $*T_i^j$, the generalization to relational quantities
is necessary. This generalization means, here, that the coordinate
axes and planes, instead of having operational meaning of momenta,
potentials, etc. characterize relations of higher order, that are
meaningful in life phenomena as well.

  The selection and determination of the relations of higher
orders is performed by a series of operations of the operator
INFORM on a set of generalized (non-dimensional) variables X_i, Y_i
and parameters P_i. The set $\{X_i, Y_i, P_i\}$ includes variables and
parameters of electromagnetic measurements, bioelectric potentials,
haemodynamic variables, etc. The operator INFORM (KRYSPIN 1969)
performs the following operations: 1) determination of independ-
ence of all generalized variables and parameters $\{X_i, Y_i, P_i\}$;

2) calculation of marginal and joint entropies of $\{X_i, Y_i, P_i\}$;
3) establishment of the relations of the second order (of the type
X_i/Y_i, X_i/P_i, Y_i/P_i); 4) establishment of the relations of higher
orders (of the type $\Pi(X_i, Y_i, P_i)$ where Π stands for generalized
products; 5) selection of the most significant relationships as
physical components of the tensor $*T_i^j$. In this way, the fundamental analytical equation of the total field is

$$*T_i^j = \text{INFORM} \times \{X_i, Y_i, P_i\}$$

Description of Experimental System. A block diagram of the
apparatus used in the collection and analysis of data is presented
in Figure 1. Data of three types is considered - the spontaneous
bioelectrical activity of the organism, the electrical resistance
of the tissue involved and the phase response to external signals.
The spontaneous bioelectric activity is separated into its con-
stituent frequency components. These are initially arbitrarily
chosen. A constant current pulse source applied in combination
with an averaging computer provides an indication of tissue resistance. A phase detector determines phase response. This data is
synchronously collected and digitized (PDP-8). Subsequently the
data is transferred in bulk to a large digital processor (IBM 7094)
at which time the generalized variables are determined and the
relevant information is calculated. These are used to modify the
data collection process for optimal results and to calculate the
tensor components.

Discussion. This approach connects the mathematical model
with the physical reality in the following way: 1) only those
quantities that can be defined operationally and measured or observed as phenomena on living objects enter as the physical components of the tensor; 2) the total tensor $*T_i^j$ is the ultimate
abstract representation of all field properties including electromagnetic fields, gravitational fields, quantum fields, etc.;
3) the operator INFORM provides a general ordering of coordination
rules according to the quantity of information contained in the
relationships between generalized variables and parameters; 4)
the physical geometry of all nature phenomena including life phenomena is characterized by a metric tensor g_{ik}, i.e. the physical
geometry is coordinate independent and operationally consistent:
the elementary length interval

$$ds^2 = g_{ik} \, dq^i \, dq^k$$

is defined in terms of operational definitions of measurements
actually performed, using generalized coordinates q^i and q^k and

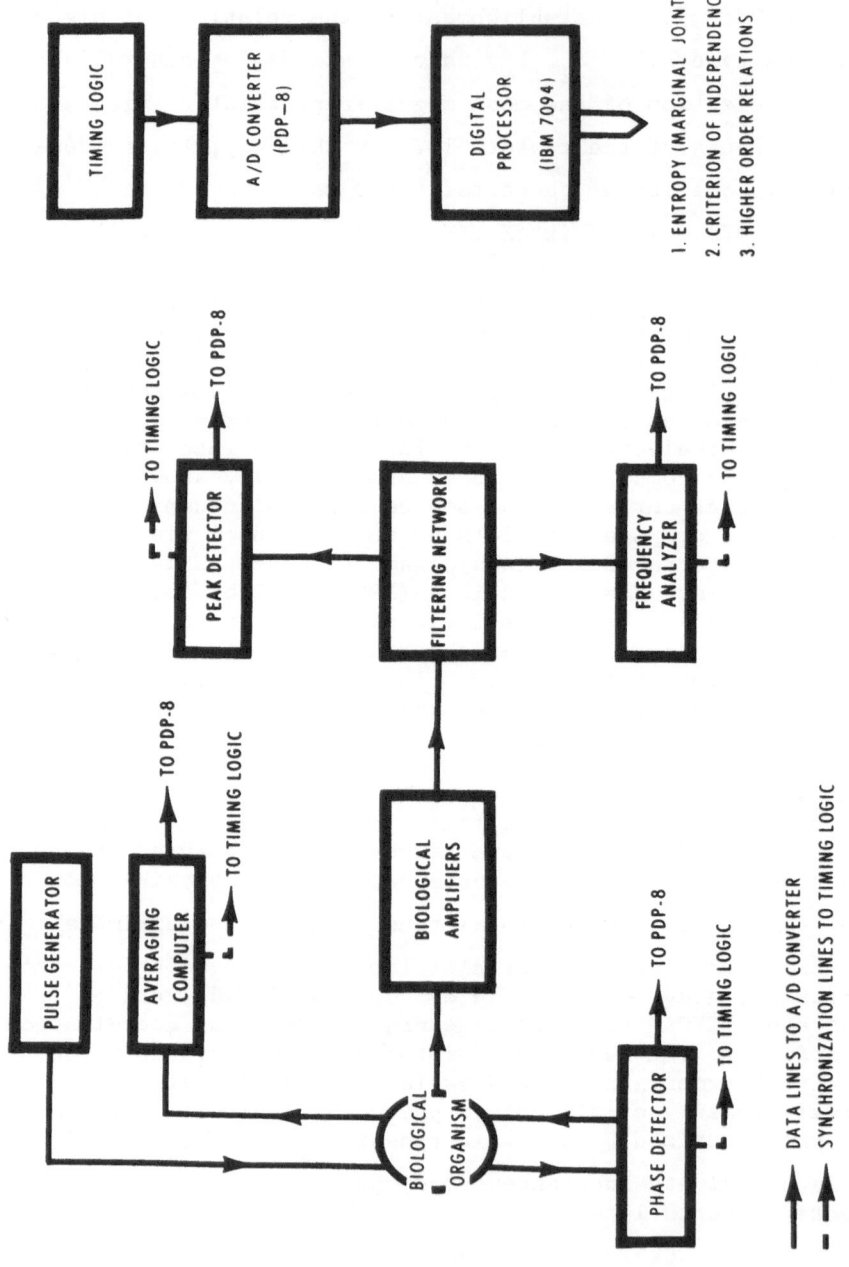

FIGURE 1.   BLOCK DIAGRAM OF EXPERIMENTAL SYSTEM

the metric tensor

$$g_{ik} = \vec{e}_i \cdot \vec{e}_k$$

(KRYSPIN 1969):　thus no transfer of coordination rules between different biological levels is allowed; 5) the use of the RIEMANNIAN geometry of a four-dimensional curved space as an ultimate mathematical representation of the order in nature limits the number of necessary physical components to 10.　This means that any given situation is always characterized sufficiently by 10 physical components and the problem is reduced to an adequate selection of 10 coordinates according to the information conveyed by them.

There is not yet enough experience with this type of description of nature, so that our understanding of the meaning of physical components of the total tensor is rather vague.　The 10 components of the tensor represent the analytical quantities of a general theory.　In the non-living physical world, they have been named densities, momenta, potentials, etc.　In the whole of nature, they must have more general meaning (EDDINGTON 1923) and in living systems they probably represent groups of relations that need not necessarily be the same throughout the whole process of observation.　The coordinates of the total field are probabilistic and their importance may change during the course of an experiment or observation during the process.

The total tensor is an invariant and may be considered as a representation of order and causality in the world.　The implications of this approach for bridging the gaps between different observation levels in biology should be obvious.

Bibliography

EDDINGTON, A.S.:　The Mathematical Theory of Relativity; Cambridge, Univ. Press, 1923

KRYSPIN, J.:　The Derivation of the Volume Conductor Geometry; in Proc. of the lVth Neuroelectric Conference, San Francisco, 1969; Charles C. Thomas Publishers, in Press

RAINICH, Y.G.:　Mathematics of Relativity; New York, John Wiley & Sons, 1950.

MAGNETIC FIELDS ASSOCIATED WITH NERVOUS CONDUCTION

Albert M. Cook and Francis M. Long

Department of Electrical Engineering: Bioengineering

University of Wyoming, Laramie, Wyoming 82070

A magnetic field arises from the motion of charge and an electric field arises from the existence of charge. Detection of magnetic fields produced by ionic currents in living tissue as well as the better known electric fields has been reported in the case of the heart and the electroencephalogram (2,3). For nerves, the electric field component associated with conduction in peripheral myelinated nerves has apparently been reliably measured (1). In addition, reports of detection of the magnetic field component accompanying peripheral myelinated conduction have been published (4,5,8). However, serious objections may be (and have been) raised regarding these latter reports on the basis that electrostatic shielding apparently was not used. The detected signal, therefore, may well have been the electrostatic component rather than the magnetic component. This paper discusses the results of an investigation which attempted to measure the magnetic fields associated with nervous conduction.

In order to obtain an approximate value for the theoretical magnetic field existing outside a conducting axon, it was necessary to examine the currents which exist during conduction. There are two primary types of currents which play important roles in nervous excitation (6). The first of these is the ionic currents which are due to the movement of ions across the axonal membrane. The second is the longitudinal currents which are generally held to be responsible for the propagation of the nervous impulse. Of these two types of current, the longitudinal currents will most likely give rise to external magnetic fields which would be detectable using probes placed parallel to the conducting nerve. The longitudinal current reaches a maximum value of approximately 2.0×10^{-9} amperes

at a node of Ranvier, and has a time course similar in shape to a
rectangular pulse of 0.25 msec duration (7).

In order to apply electromagnetic field theory to nervous conduction, the longitudinal current will be considered to be a rectangular pulse of 2.0 X 10^{-9} amperes amplitude and 0.25 msec in
duration. The mathematical relationship between the current density,
$\overline{J}$, and the magnetic field intensity, $\overline{H}$, for the time-varying case
is given by Equation (1).

$$\nabla \times \overline{H} = \overline{J} + \frac{\partial(\varepsilon\overline{E})}{\partial t} \qquad amp/m^2 \tag{1}$$

where $\overline{E}$ is the electric field intensity and ε is the dielectric constant of the medium. From the work of Burr and Mauro, it is known
that the electric field associated with the nervous impulse has a
very slow variation with time (1). Since ε most likely is constant
with time, Equation (1) may be closely approximated by the static
form of Ampere's law given in its integral form by Equation (2).

$$\oint \overline{H}\cdot dl = I \qquad amp \tag{2}$$

Integrating Equation (2) around any closed path which includes
the longitudinal current yields the following result.

$$H = \frac{I}{2\pi r} \qquad amp\text{-}turn/m \tag{3}$$

Since the detector employed in this study was placed at approximately 1 mm from the conducting nerve, r will be taken to be 1 mm.
Substituting these values into Equation (3), the value obtained for
H is 3.19 X 10^{-10} amp-turns/meter or 4.0 X 10^{-9} oersteds. This
value compares favorably with Seipel's estimated value of 12.0 X
10^{-9} oersteds (8).

This extremely low estimated value for H would be quite difficult to measure and, therefore, it was necessary to employ magnetic
detectors of great sensitivity. Two basic types of detectors were
employed in these experiments. The first type consisted of two
specially designed magnetometers. The second type was a commercially
available magnetic tape recording head consisting of a large number
of turns of wire machine wound on a high permeability core and surrounded by a shielding material. The need for shielding was elimi-
nated in the case of the magnetometer circuits by the use of a nulling procedure which resulted in a balancing of the circuits in the
presence of any ambient fields. In both of these circuits, a high
permeability toroid core was used as a magnetic detector and electronic amplification with proper bandwidths was used prior to recording or displaying. The output of the tape head was filtered and
then amplified before being displayed on an oscilloscope.

The experimental apparatus is shown schematically in Figure 1.
The sciatic nerve preparation from Rana pipiens was placed in a

chamber equipped with four electrodes. The magnetic field detectors
were mounted on an adjustable probe. In Figure 1, the block labeled
"probe" refers to the toroidal core in the case of the magnetometer
circuits or to the tape head. The block labeled "added circuitry"
refers to either the magnetometer circuitry or to a filter and preamplifier in the case of the tape head experiments. A dual beam
oscilloscope was used with the magnetic activity displayed on one
channel and the electrical activity displayed on the other. Permanent records were obtained using a Polaroid camera.

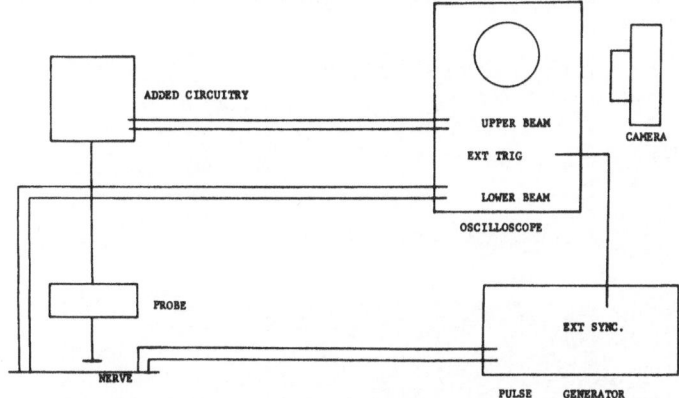

Figure 1: Block diagram of experimental apparatus

In order to evaluate the performance of the detectors, a wire
carrying a current pulse of less than 2.0×10^{-9} amperes was sub-
stituted for the nerve preparation and the output of the detector
was monitored. Figure 2 is a composite picture showing the output
of each of the detectors used in this study. From this figure it
may been seen that all three detectors had a measurable output for
the applied current.

After determining that the magnetic detectors were sensitive
to fields from currents similar in magnitude to the external longitudinal currents reported to exist during conduction, detection of
related magnetic fields was attempted in more than 30 sciatic nerve
preparations from Rana pipiens. Using the experimental apparatus
illustrated in Figure 1, the nerve was placed in the chamber and
the electrical activity was monitored to insure that the nerve was
responding to the applied stimulus. When satisfactory electrical
activity was measured, the probe was lowered to approximately 1 mm
from the axon and the output of the magnetic detector was observed.
A typical record obtained from these experiments is shown in Figure 3. In this figure, the magnetic activity is shown on the upper
trace and the electrical activity is shown on the lower trace.
Careful examination of the upper trace reveals that the only magnetic signal which appears is that due to the stimulus artifact.

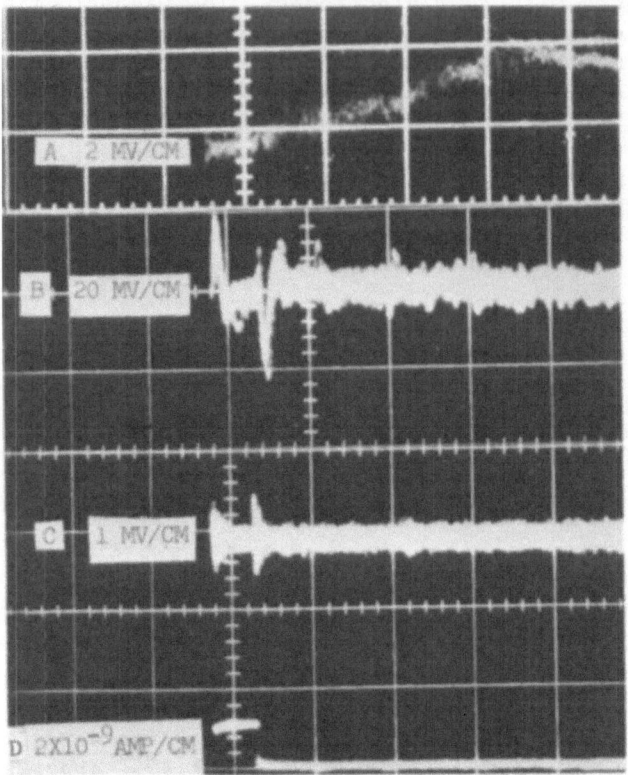

Figure 2: Composite photograph showing outputs of magnetic detectors for current shown in D. A and C: outputs of magnetometers, B: output of tape head.

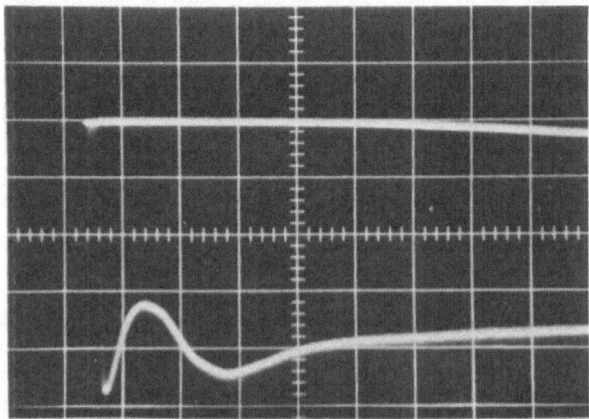

Figure 3: Typical record from nerve experiments. Upper: magnetic activity. Lower: electrical activity. Both vertical scales: 20 mv/cm, horizontal scale: 1 msec/cm.

In each experiment carried out with responding nerves, the magnetic detector showed no signal which could be associated with the propagating action potential. Control experiments were also performed with non-responding nerves and with a silk thread soaked in Ringer's solution. In these runs, the magnetic output was essentially the same as that shown in Figure 3.

The results obtained were evaluated in relation to the accepted local circuit theory. According to this theory, the longitudinal currents exist in small local loops with the measured outer current representing only a part of the total effect (6). A current of the same magnitude but opposite sense exists inside the fiber. Since the two currents are separated by only about 100 angstroms (the approximate thickness of the axonal membrane), the field arising from the inner current can be calculated using r=1.001 mm. This value is 3.99×10^{-9} oersteds. Thus the estimated value of the resultant magnetic field would be the difference between the values from the outer and inner currents or about 10^{-11} oersteds. Detection of a field of this size would require detectors with sensitivities two orders of magnitude greater than those employed in this study.

In view of the data presented herein, it would appear that an indirect verification of the local circuit theory could be inferred. It can also be stated with some certainty that the experiments performed as described herein have raised a question as to the validity of the claims regarding detection of the magnetic field associated with a propagating action potential.

References

1. Burr, H.S. and A. Mauro, Electrostatic fields of the sciatic nerve in the frog, Yale Biol. Med., 21, 455, 1949.
2. Cohen, D., Magnetic fields around the torso: production by electrical activity of the human heart. Science, 156, 652, 1967.
3. Cohen, D., Magnetoencephalography: evidence of magnetic fields produced by alpha-rhythm currents, Science, 161, 784, 1968.
4. Gingerelli, J.A., N.J. Helter and W.R. Glasscock, Magnetic fields accompanying transmission of nerve impulses in the frog's sciatic, J. Psychol, 52, 317, 1961.
5. _____, Further observations of magnetic fields accompanying nervous transmission and tetanus, J. Psychol., 57, 201, 1964.
6. Hodgkin, A.L., The Conduction of the Nervous Impulse, Springfield, Illinois: C. C. Thomas, 1964.
7. Huxley, A.F. and R. Stampfi, Evidence for saltatory conduction in peripheral myelinated nerve fiber, J. Physiol., 108, 315, 1949.
8. Seipel, J.H. and R.D. Morrow, The magnetic accompanying neuronal activity, J. Wash. Acad. Sci., 50, 1, 1960.
9. Seipel, J.H., Personal correspondence.

CURRENT DENSITY MEASUREMENT - AN ATTEMPT FOR IMPROVEMENT

H. J. Marsoner, M.D.

Chirurgische Universitaetsklinik, Graz, Austria

The problem of the current density distribution in the brain during the application of diffuse electric currents for electro-anaesthesia and electrotherapy has been discussed during the last few years.

Some of the first measurements were carried out by CLUTTON BROCK and presented in 1966 at the First Graz Symposium. At the same time MARSONER et al. discussed some problems of electrode design. Later on TATSUNO et al. (1968) published their design, construction and application of a probe for current density measurement.

All these attempts are characterized by the use of a bipolar probe which is connected to a current measuring device with a very small internal resistance. The probes were calibrated in terms of current density, but can detect only one of its components.

Nevertheless, there is another source of error to which most of the people using current detecting probes did not pay enough attention. If a probe is calibrated in a medium of a certain specific resistivity, it cannot be concluded that the calibration factor is the same when the electrode is used in a medium of a different conductivity.

It is not necessary to discuss why a bipolar probe, connected to a low input impedance current measurement device, cannot correctly detect the current density in an nonhomogeneous medium like the brain, where the specific resistivity changes considerably from site to site, as STEINKE (1958) has shown. This was already discussed extensively by BAKER at the Second Graz Symposium. As a result of the condiseration of the field theory we find that a bipolar

15

probe can only detect a voltage difference. All the attempts to
use a load resistor between the two electrodes of the probe give
incorrect or at best approximate values of the current density.

     Therefore, we have to use another way. My concept was the
following: In a particular site of measurement the specific resistivity of the brain tissue and the voltage drop at the two electrodes of the probe are measured. Then the probe is calibrated in
a homogeneous current flow field using electrolyte solution with
different specific resistivites. Proceeding that way, a relation
is obtained between the original current density of the undistrubed
field and the voltage drop across the bipolar probe. The parameter
is the specific resistivity of the calibration-medium. In other
words, we have a calibration factor for any particular value of
specific resistivity of the brain.

     I found out, that the probe described by DEUTSCH is not only
useful for detecting three orthogonal components of the current
density vector, but also for the measurement of the specific resistivity of the tissue. As you surely remember, DEUTSCH's probe is
a tetrapolar probe. Although the single electrodes of the probe
are placed in the corners of a square it can be used for four-ter-
minal measurements of the resistivity. This has the main advantage
that polarization-effects are cancelled out because the voltage detecting pair of electrodes is connected to an amplifier of very
high input impedance. Theoretical considerations on the field distribution around the probe show, that only a few cubic millimeters
of tissue are concerned in the measurement. This has an important
advantage compared with usually applied techniques for in situ measurement (STEINKE). The probes used for the above mentioned techniques usually integrate over a wide area of tissue, whereas the
very thin probe used in these experiments detects a more local value.
This explains also that the values measured by us are close to the
maximum values reported in literature.

     Systematic measurements were carried out in our laboratory on
five cats in different anterio-posterior planes. In each plane the
electrode was introduced on five different lateral positions. Beginning a few millimeters under the cortical surface a value was
taken every two millimeters. The electrode was inserted using
stereotaxic techniques.

     The current supply electrodes were implanted and attached to
the surface of the skull frontally and occipitally at the level of
the zero plane. Sinusoidal current of 1000 cps and an intensity
of 2.5 mA was applied. The measurement of the resistivity of the
tissue was also carried out using a constant 1000 cps sinusoidal
current.

     As already reported by NEIDINGER we find a very nonhomogeneous

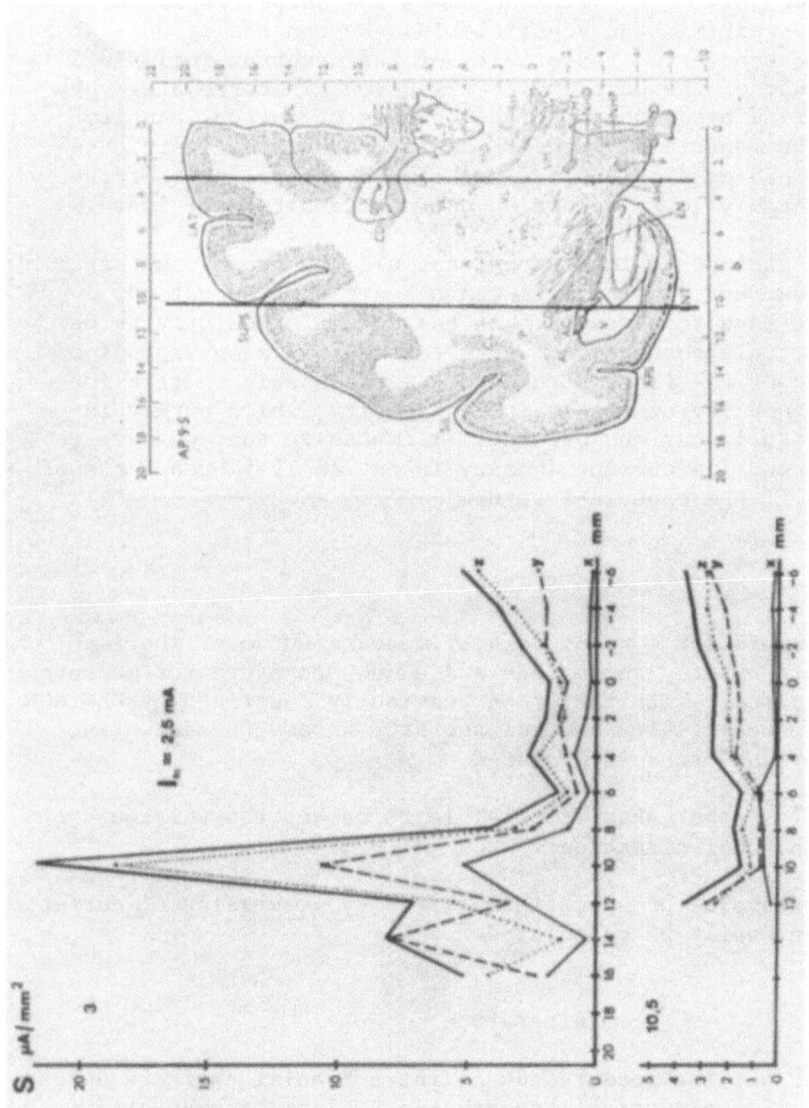

Figure 1

(a) Upper and lower graphs show the current densities at 3mm. and 10.5 mm respectively lateral from the media. (b) Coronal plane AP 9.5 showing sites of measurement.

distribution of the current density. The figure shows one example among our measurements. The plane AP 9.5 was selected for demonstration. (Figure 1)

The upper part of the diagram shows how surprisingly large the current density in the ventricle is. We can easily understand this finding because the resistivity of the cerebrospinal fluid is a tenth of that of the surrounding tissue. This correlates very well to our findings concerning the changes of current density near the third ventricle, when it was filled by a dielectric substance (MARSONER et al. 1969). In the more lateral part of the brain considerably lower values of current density were found.

In this attempt at improvement not all the questions that arise when current density measurements are carried out could be answered. We have to remember that the specific resistivity of the brain not only changes from site to site but also depends on the direction of the fibre structure. I was unable to take into account this quality of the brain resistivity, which surely introduces errors in our measurement. I emphasize that we have to be careful measuring current density in extremely fibred parts of the brain, e.g. the capsula interna.

SUMMARY

The attempts for current density measurement over the last few years are briefly summarized and a new procedure for measurement is described. With the probe previously described by DEUTSCH both specific resistivity and voltage drop across the detecting electrodes of the probe is measured.

DEUTSCH's probe makes it possible to detect the voltage drop in three orthogonal directions.

Using the value of specific resistivity, the value of current intensity can easily be calculated.

Literature

J. Clutton-Brock: The measurement of intra-cranial currents during electrical anaesthesia. Electrotherapeutic Sleep and Electroanaesthesia. Proceedings of the First International Symposium, Graz, Austria, September 1966.
J. Tatsuno, R.L. Zouhar, R.H. Smith and St. C. Cullen: A probe for determining current magnitude and pathway in the brain. Medical Research Engineering, volume 7, number 1, 1968.

H.J. Steinke and W. Buchholz: Operative Leitfahigkeitsbestimmungen
                    des Hirngewebes zur Ortsdiagnostik raumfordernder
                    Prozesse. Acta Neurochirurgica, Suppl. VI, 1959,
                    page 9.
L.E. Baker and L.A. Geddes: The measurement of current density dis-
                    tribution in biological materials. Proceedings of
                    the Second International Symposium on Electrosleep
                    and Electroanaesthesia. In print.
S. Deutsch: A Probe to monitor electroanaesthesia current density.
                    IEEE Transactions on Bio-Medical Engineering, vol-
                    ume BME-15, No. 2, 1968, page 130.

A METHOD OF EVALUATING ELECTRODE PLACEMENTS USED IN

IMPEDENCE CEPHALOGRAPHY

Daniel A. Driscoll

Union College

Schenectady, N.Y.

In previous reports to this conference, a theoretical model
of the human head was presented; the <u>direct</u> interpretation of this
model [1] was used to relate currents applied to the surface of
the head to the resulting current densities in the brain, while
the <u>reciprocal</u> interpretation [2] was used to predict the surface
potentials produced by electrical sources in the brain. In this
paper, the two interpretations are used together to evaluate the
effectiveness of various surface electrode positions for measuring
impedance fluctuations in the brain.

INTRODUCTION

In impedance cephalography [3] (the term is intended to include
rheoencephalography and plethysmography) the usual procedure is to
monitor the impedance between two electrodes on the surface of the
head; a constant current is applied to the electrodes, and the resulting potential fluctuations, which are proportional to the impedance fluctuations, are taken as an indication of cerebral blood
flow or of blood volume changes.

A more detailed analysis of the procedure considers the application of current and the measurement of the potentials as separate
functions, as if two sets of electrodes were used. A small current
is applied to one pair of electrodes on the surface of the head.
The resulting current density in the brain at the origin of the
impedance fluctuation produces a fluctuating potential across that
impedance, and a second set of surface electrodes (or in most cases
the same set) is then used to record the surface potentials resulting
from the fluctuating potential in the brain.

CRITERIA FOR ELECTRODE PLACEMENTS

The placement of two sets of electrodes should therefore at least be considered: one set to supply current to the region of the impedance fluctuation, therby producing a potential fluctuation, and a second set to measure the resulting potential fluctuation on the surface of the head. It is desirable that the current electrodes supply as much current as possible to the region of the impedance fluctuation to be measured so that the greatest possible potential fluctuation is produced; at the same time it is desirable to supply as little current as possible to other regions where potentials from similar impedance fluctuations may contaminate the desired signal.

A similar criterion applies to the potential measuring electrodes which should be maximally sensitive to potentials from the region of the impedance fluctuation of interest, and minimally sensitive to potentials resulting from other impedance fluctuations. In terms of the reciprocity theorem [4] these criteria for the potential and current electrodes are identical and always result (theoretically) in identical positions for the two sets of electrodes. Practically, there are other factors such as electrode size which restrict the choice of electrode positions; in some situations, therefore, separate positions for the current and potential electrodes may be indicated.

MODELING THE IMPEDANCE FLUCTUATION

For the purpose of this discussion, consider that the impedance fluctuations due to pulsations in the diameter of a blood vessel lying in the cerebral cortex are to be measured. The blood vessel must first be modeled for quantitative considerations. In the very simple model shown in Figure 1, it is assumed that the blood vessel is a long homogeneous circular cylinder of radius "a" and resistivity ρ_a surrounded by a homogeneous brain of resistivity ρ_b. (A more complete model including the vessel wall and the axial accumulation of red cells could just as easily be used.) For maximum response, the surface current electrodes are positioned so that current flow is normal to the axis of the blood vessel, and the additional assumption is made that the resulting field (E) in the neighborhood of the blood vessel is uniform except for the perturbation field caused by the blood vessel which is:

$$V = \left[2\pi E\, a^2\, \frac{(\rho_a - \rho_b)}{(\rho_a + \rho_b)} \right] \frac{\cos\theta}{2\pi r} \quad \text{Volts} \qquad (1)$$

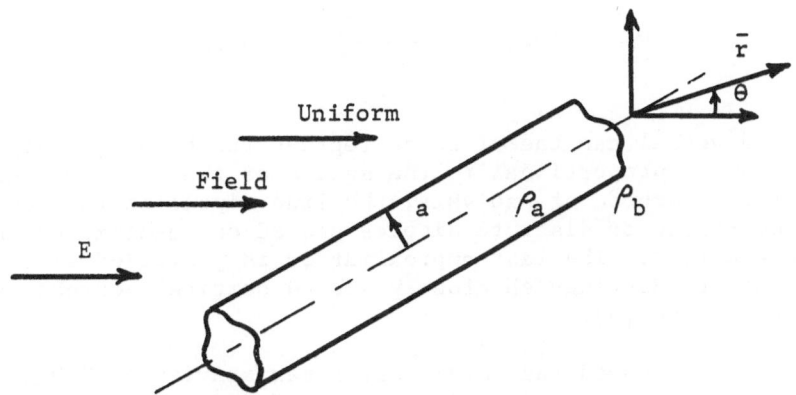

Figure 1. A simplified model of a
blood vessel in a homogeneous
brain with uniform applied field.

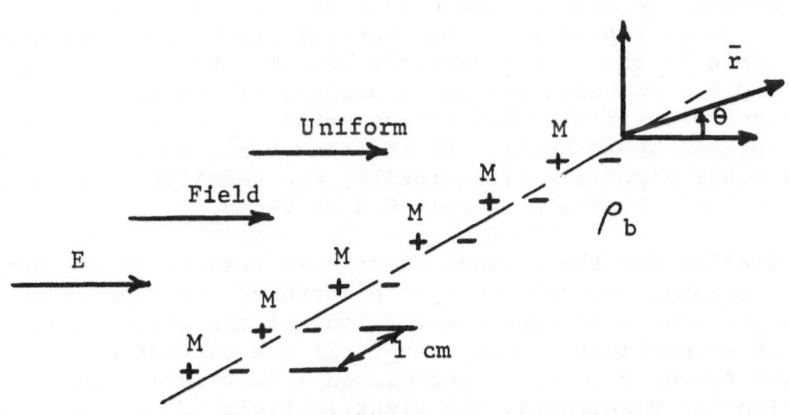

Figure 2. Discrete dipole model of
the blood vessel.

This is the same field as would be produced in a homogeneous medium
of resistivity ρ_b by a line dipole of strength

$$M_L = 2\pi E \, a^2 \frac{(\rho_a - \rho_b)}{(\rho_a + \rho_b)} \qquad \text{V--cm} \qquad (2)$$

The blood vessel can therefore be represented by a line dipole
whose magnitude is proportional to the square of the blood vessel
radius. For the purpose of analysis, the line dipole is then approximated by a line of discrete dipoles spaced one centimeter apart
as shown in Figure 2. The last approximation is justified since
it is difficult to distinguish closely spaced cortical sources using
surface measurements [5].

If it is now assumed that brain has a resistivity of 200 ohm-cm
and that the blood vessel has a resistivity of 300 ohm-cm, then the
resulting magnitude of each dipole moment is

$$M = 2\pi \frac{100}{500} E \, a^2 = 1.3 \, E \, a^2 \quad \text{V--cm}^2 \qquad (3)$$

AN ELECTRODE PLACEMENT FOR CORTICAL MEASUREMENTS

A satisfactory placement for the potential measuring electrodes
(particularly if signal averaging techniques are used) is directly
over the blood vessel, with an interelectrode spacing of one centimeter, and having the line of the electrodes at right angles to the
length of the blood vessel [5]. The sensitivity of this electrode
pair to a dipole in the cortex directly beneath the midpoint of the
electrodes and 0.2 cm below the inner surface of the skull is 4.6
$mV/V\text{-cm}^2$ (from theoretical calculations using a concentric spheres
model of a typical human head). At the same time, these electrodes
reject most other signals satisfactorily; the sensitivity to a source
near the center of the brain is only 0.3 $mV/V\text{-cm}^2$.

The criterion for the current electrodes results in the same
electrode positions, however, current electrodes are usually of
such a size (to reduce voltage fluctuations at the electrode-scalp
interface) as to prohibit 1 cm spacing. If the current electrodes
are therefore spaced 5 cm apart and placed symmetrically in line
with the potential electrodes, the electric field (E) produced in
the neighborhood of the blood vessel will be normal to the axis of
the blood vessel and about 2 mV/cm per mA of applied current for
the typical human head, while the field near the center of the brain
is only about 0.2 mV/cm per mA.

Considering just one of the line of dipoles used to model the
blood vessel (the dipole beneath the midpoint of the electrodes),
the dipole strength is (from Equation 3) 1.3 E a^2 V-cm, where

(from the previous paragraph) E = 2 mV/cm per mA applied, or M = 2.6 a^2 mV-cm/mA. Using the sensitivity of the potential electrodes given above, the surface potential measured is then (4.6 mV/V-cm)* (2.6 a^2 mV-cm/mA) = 0.012 a^2 mV/mA. If the blood vessel radius varies from a = 0.10 to 0.12 cm, the resulting peak-to-peak voltage fluctuation per mA of applied current is 0.012(0.0144-0.0100) = 53*10^{-6} mV/mA. This term has the units of impedance and is usually interpreted as an impedance fluctuation when the same electrode pair is used to apply the current as well as measure the potential.

It can be calculated from other data given above that the response to pulsations of a similar blood vessel near the center of the brain would be about 150 times less than the response calculated for the cortical blood vessel, (or if another model were made, the response to changes in size of a cerebral ventrical could be calculated). An impedance fluctuation near the center of the brain would therefore contribute little to the surface potential measurement using this particular electrode system.

In a practical situation it would be necessary to consider several more of the dipoles used in modeling the blood vessel, and by superposition obtain a better estimate of the resulting potential. Consideration of additional impedance fluctuations in the scalp as well as in the brain (including cerebral ventricles) would also be necessary to estimate the degree to which the desired signal is contaminated by unwanted signals.

In this paper, a method is presented for evaluating electrode placements used in impedance cephalography; the method makes use of a previously reported theoretical model of the human head. While the example presented does not accurately reflect the complex considerations necessary in impedance cephalography, it serves to demonstrate a technique which can be used to obtain a more quantitative evaluation of the electrode systems than was previously available.

REFERENCES

[1] Rush, S. and Driscoll, D.A., "A mathematical model for current flow in the brain from surface electrodes," Abstracts of the Neuroelectric Conference, 4: 11; Milwaukee, October, 1967.

[2] Rush, S. and Driscoll, D.A., "A model of the head for calculating EEG electrode sensitivity," The Neuroelectric Conference; San Francisco, February, 1969.

[3] Lifshitz, K., "Electrical Impedance Cephalography," Manuscript; Research Center, Rockland State Hospital, Orangeburg, N.Y.; 1967.

[4] Rush, S. and Driscoll, D.A., "EEG electrode sensitivity-an
     application of reciprocity," IEEE Transactions on Bio-medical
     Engineering, BME 16,1: 15-22; January, 1969.

[5] Driscoll, D.A., "A theoretical analysis of several electrode
     placements on the scalp for recording averaged evoded potentials,"
     Proceedings of the Neuroelectric Conference; March, 1970.

AN INVESTIGATION OF EXCITABILITY THRESHOLDS USING TWO CLOSELY-SPACED HIGH-FREQUENCY SINUSOIDS

(Supported by PHS Grants 1 F03 HE-44525, 1 T1 HE 5875, HE 09381-04)

Stanley Rush, Ph.D., and J. A. Abildskov, M.D.

University of Vermont, E.E. Department, and

University of Utah, Department of Internal Medicine

INTRODUCTION

Diverse waveforms and focusing methods have been employed to induce electrical anesthesia and convulsive shock with the hope of improving the safety and efficiency of these procedures. Recently, two repetitive waveforms closely spaced in frequency and introduced through two pairs of electrodes have been reported as having some desirable properties in clinical studies.[1]   A logical framework for this approach is schematized in Fig. 1. Assume current waveform $I_1(t)$ is applied to the head through electrodes a and b and $I_2(t)$ is similarly applied through electrodes c and d. Near a and b the current mainly reflects waveform $I_1(t)$ and near c and d $I_2(t)$. At some point, e, in the center of the head $I_1(t)$

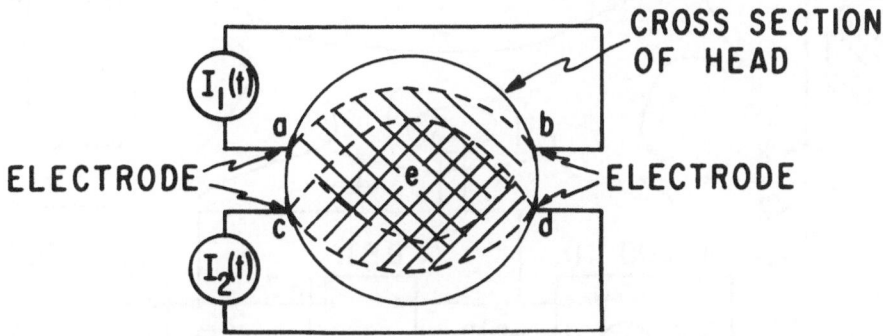

Fig. 1   Schematic of current flow from two pairs of electrodes. Shaded areas show regions of high current intensity from each pair; crosshatched where they add with approximately equal amplitude.

and $I_2(t)$ sum to give a third waveform $I_3(t)$ that may reach a
critical level for stimulation $I_3^c(t)$.

If $I_1(t)$ and $I_2(t)$ vary in time in exactly the same way, $I_3(t)$
also will vary in that fashion. Assume in this case that when the
critical level $I_3^c(t)$ is reached there is an unwanted side effect
(such as scalp muscle stimulation) produced immediately under the
electrodes. As a possible technique for improving this method, we
may attempt to utilize two <u>different</u> time-courses for $I_1(t)$ and
$I_2(t)$ such that when $I_3^c(t)$ is reached, the current intensities
directly under the electrodes have not yet reached levels at which
the side effects appear.

In the previous publications, $I_1(t)$ and $I_2(t)$ were relatively
high frequency sinusoids or square waves which differed in their
repetition rates by a few hundred cycles per second. The investigation reported here was an attempt to verify by quantitative laboratory measurements the existence of such a synergistic effect
and, if present, to obtain evidence of its mechanism.

METHOD

In this study, the dog heart was utilized as the excitable
tissue. The entire heart responds in an all-or-none manner to
threshold levels of current density and the QRS which results from
stimulation of the ventricle is recognizably different from that of
a normally conducted complex. The left side of the heart of an
anesthetized dog was exposed in a conventional preparation, Fig. 2.
The ECG from body surface leads and the detected difference frequency were displayed and recorded as shown.

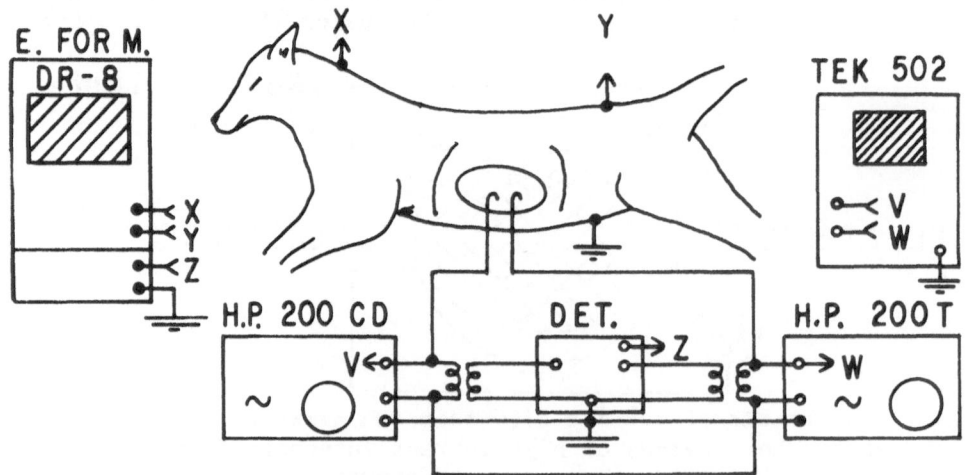

Fig. 2 Circuit for double-sine-wave threshold experiment.

The signals usually were connected to the heart through a single pair of small hooked electrodes in the myocardium. (In one set of experiments two pairs of such electrodes were used with one oscillator connected to each.) The applied voltage was monitored for peak-to-peak amplitude by a separate oscilloscope. The difference frequency was obtained from a simple diode circuit especially designed for the purpose.

Several studies were carried out with varying degrees of detail. The first investigation used a single oscillator and measured thresholds over a range of frequencies from 1 hz (a H.P. 3310A function generator was used at this frequency) to 20,000 hz. An occasional beat induced by the electrical stimulus was taken as an indication that threshold voltage had been reached. It was, however, possible to capture the heart rate at a submultiple of the oscillator frequency at frequencies up to 500 hz. The second investigation used double-frequency stimulation with equal components from each oscillator and thresholds were measured only when the heart rate was completely locked to some submultiple (typically 1/3 to 1/8) of the beat (difference) frequency. Other experiments considered the variation of threshold with beat frequency at given high (carrier) frequencies and the effectiveness of using combinations of different oscillator voltages as compared to equal oscillator voltages.

RESULTS AND DISCUSSION

The literature contains data from numerous studies of the stimulus thresholds of excitable tissue. These typically utilize electrical pulses of varying lengths and strengths and the relationships are usually displayed on strength-duration curves.[2] The initial investigation of frequency vs. threshold was carried out in order to determine whether, in the heart, the relation between sinusoidal waveform stimulation and pulse stimulation follows the classical interpretation. The flat (rheobase) portion of the strength-duration curve is due to accommodation; an effect negligible where the curve rises steeply with decreasing pulse lengths. A sine-wave interpretation considers each half-cycle as a pulse. At high and medium frequencies a plot of half-cycle lengths vs. strength should appear like the strength duration curve; the rise times of the sine waves not being significantly different from those of rectangular pulses. At low frequencies, the slope of the sine-wave voltage (and current) is significantly different from the rapid initial rise of a rectangular pulse and permits accommodation to take place before threshold is reached. As a result, the sine wave signals require more amplitude to stimulate at very long half-cycle lengths than at intermediate values. Correspondingly, the results of this study show a rapidly rising threshold with higher frequencies, a flat region between 300 and 30 hz and a steady rise below 30 hz to a threshold at 1 hz about 3 times the rheobase level.

The second and principal investigation measured the synergistic effect of utilizing two frequencies as compared to one. The ex-
pected pattern of two closely-spaced sinusoidal frequencies of equal
amplitude is shown in Fig. 3.

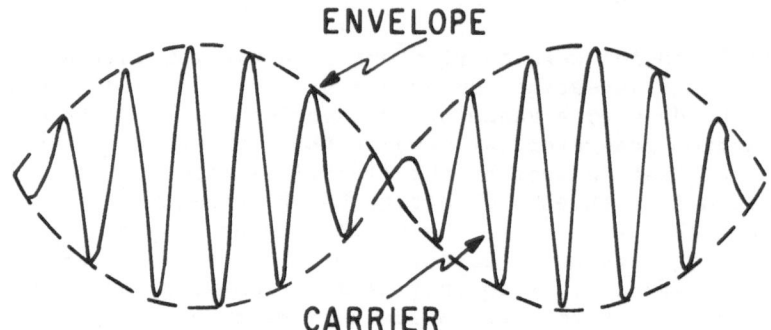

Fig. 3 Waveform of double frequency current; carrier frequency
twelve times beat frequency.

The upper and lower parts of the envelope have the characteristics
of the "full-wave-rectified" difference frequency while the oscillations themselves are at the mean (carrier frequency) of the two
high frequencies. The maximum peak-to-peak voltage at threshold
for double-frequency stimulation was compared with peak-to-peak
voltage thresholds for single frequency stimulation over a range of
carriers from about 300 to 15,000 hz. Each set of curves had the
features associated with a strength-duration curve but, as shown in
Table I, less voltage was required at frequencies above 500 hz with
the double-frequency method.

TABLE I

f (hz)	320	640	1,280	2,500	5,000	8,000	10,000
R	.9	3.2	3.1	2.4	2.6	1.9	1.7

R is the mean ratio (three measurements) of threshold voltage
using a single frequency, f, to the maximum at threshold of the
sum of two equal sinusoids at slightly different frequencies. The
mean ratio (640 f 10,000) is 2.5:1. The data are from one dog using
3 pairs of electrode sites, but preliminary experiments on three
other dogs showed similar results. The detected envelope and the
driven and normal QRS are shown in Fig. 4.

In two experiments four electrodes in the heart were placed to
simulate the arrangement of Fig. 1. The results suggested that
stimulation did not take place at point, e, where the currents were
equal. As a check, thresholds were measured as in Fig. 2 but with

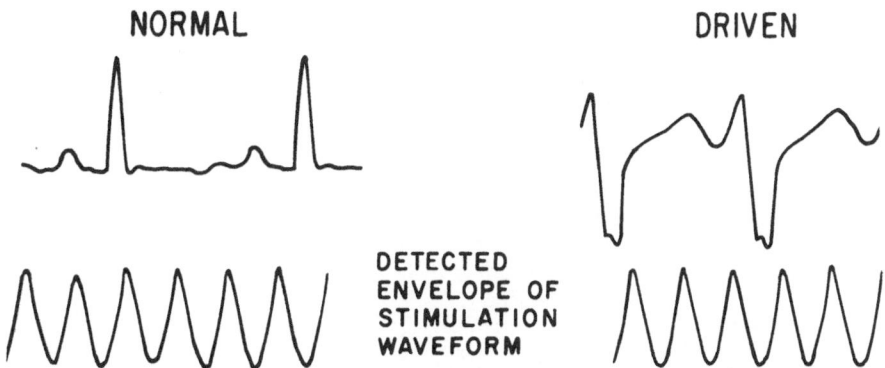

Fig. 4 Example of recorded waveforms; normal ECG upper left, driven ECG upper right and detected beat frequency below.

various combinations of oscillator voltages. Only a small amount of data was taken but it showed that the same peak-to-peak voltage was effective as a stimulus for ratios of the individual voltages as high as 2:1.

In the last study, a comparison of peak-to-peak threshold voltages as the beat frequency was changed with the carrier remaining fixed was carried out. The data were quite variable although the trend was present in all experiments. There seemed to be a definite requirement for larger stimuli under one-to-one drive, about 3 cycle-sec. as compared to heart rate drive at submultiples of the beat frequency. In one set of measurements, the ratio of threshold at 4:1 drive ratio as compared to 1:1 at three carrier frequencies (1,280, 2,500 and 5,000 hz) averaged 1:1.2 corresponding to a twenty percent difference.

CONCLUSIONS

The main conclusion of this investigation is that there is a definite synergistic effect when two closely spaced frequencies are used as compared to a single frequency. A detailed investigation of the mechanism is of manifest importance but beyond the scope of the study being reported. A possibility which appears to fit the observed facts is that the membranes act in a non-linear fashion at high drive levels and thus tend to rectify the modulated signal, that is effectively to produce an additional component of charge transfer across the membrane at the beat frequency where lower thresholds exist.

1. Smith, R.H. et al; Abstr. of Neuroel. Conf. Vol 5, 1969, P.8.
2. Brazier, Mary, The Electrical Activity of the Nervous System, Williams and Wilkins, Balt. 1968, P.82; 3rd ed.

SPATIO-TEMPORAL POTENTIAL FIELDS OF EVOKED RESPONSES RECORDED

TRANSCRANIALLY: PRELUDE TO STIMULATION

D. G. Childers, J. R. Bourne, N. W. Perry

Electrical Engineering, University of Florida, Vanderbilt

University and Psychology, University of Florida

INTRODUCTION

The spatio-temporal potential field characteristics of the EEG have been investigated over the years by numerous investigators but there have been less investigations of averaged responses (see, e.g. Gastaut and Regis, 1965, Remond 1965). We were motivated to use an array of electrodes in our study of the visual evoked response (VER) for numerous reasons, among which are included the detection of loci for visual defects such as amblyopia and the determination of anatomical locations of information processing and perception of various visual stimuli. Thus, we seek to monitor cortical information processing occurring temporally in numerous spatial neuronal populations (ensembles). It is our belief that a study of the spatio-temporal potential field characteristics of the VER obtained as a result of monitoring an array of electrodes located on the occipital scalp of humans will provide us with insight into how the brain encodes, processes, and perceives visual stimuli.

This paper will present findings obtained as a result of examining the characteristics of spatio-temporal maps of the VER. These findings suggest an asynchronous firing of underlying neuronal ensembles and we, thus, have been led to conclude that the scalp potential appears to be actively generated by several spatial sources (Clynes and Kohn, 1967).

Several models will be advanced that explain the observed phenomena, and finally we offer a rational for the conjecture that the results of this study are a prelude to transcranial stimulation by a scalp electrode array.

THE SPATIAL VISUAL EVOKED RESPONSE
Since the VER is an evoked response measured differentially

33

T = 230MILLISECONDS

FIGURE 1
TOP HALF

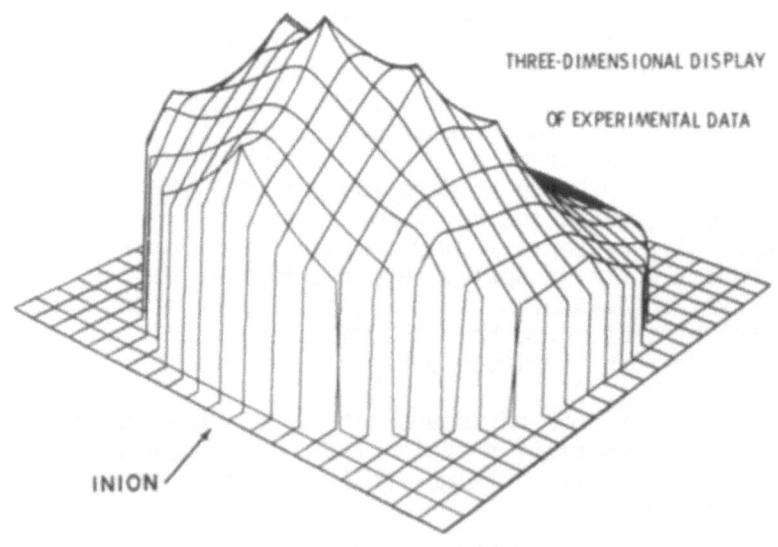

THREE-DIMENSIONAL DISPLAY

OF EXPERIMENTAL DATA

INION

between two electrodes we have chosen to call the spatio-temporal potential field characteristics of the VER the spatial visual evoked response (SVER). An extensive historical review of the various electrode arrays used in the past, the production of spatio-temporal maps, and associated models can be found in Bourne (1969).

A circular array was chosen for this study. The array consists of 17 electrodes arranged in two concentric circles of 8 electrodes each with an electrode at the center. The electrode spacings on radial lines are two centimeters. This particular electrode configuration was selected for several reasons among which are included the fact that a potential hill in the occipital region could be easily measured with this array, also, certain positions in this array are almost directly comparable with arrays used in various other experiments in other laboratories.

In order to gain a concise impression of the potential spread that occurs temporally, it is necessary to rapidly scan the spatio-temporal maps in temporal order. This is best done in the form of a movie which is presented as part of this paper. (Examples of the two types of displays that appear in the movie are shown in Figure 1).

Numerous subjects have been examined by this technique with various visual stimulation, e.g., fovial and off-fovial stimulation, white and colored light, and in addition, both monocular and binocular stimulation have been used.

Summary of Experimental Results. The most significant result of this study is that the summated evoked response obtained from a set of 16 electrode pairs located over the occipital scalp displays a characteristic spread of potential common to all subjects. This spread takes the form of a rotation of a potential which initially rotates either clockwise or counter-clockwise, forms a quasi potential hill, and finally returns to its original position. Next, there are certain characteristic features of the SVER independent of the stimulus used. At about 240 milliseconds, the highest point of the potential gradient exists in the lower half plane of the electrode array or the most posterior portion of the scalp. This feature of the SVER appears to be invariant with stimulus condition. A potential spread precedes and follows this characteristic condition. However, the rate of spread is not invarient and may differ significantly from one stimulus condition to the next.

The largest signal to noise ratios are found in the anterior portions of the electrode array. This finding would indicate that the VER recordings should be made from these electrode locations in order to obtain the best signal-to-noise ratios.

Differences between SVER's due to varying stimulus conditions are seen primarily in the early and late portions of the spatial-temporal maps. A quantative procedure has been established for determining dominance and detecing visual defects such as amblyopia. Finally, for white fovial stimulation, one can record synchronous potentials from large areas of the scalp. These recordings support the idea that the origin of the SVER is due to the firing of wide spread neuronal populations. These synchronous potentials are not spread as widely over the scalp for color data, suggesting that color stimulation may excite more specific neuronal populations than are excited by white light.

Potential Spread - Active or Passive? We have studied the scalp potential field rotations via a spatial ensemble correlation matrix procedure to determine if the potential spread is active or passive. If the observed potential were due to purely passive spread, then there would be no recurring rotation. We have observed early rotations as well as rotations at later latencies, and it is hypothesized that two generator mechanisms are responsible for these separate phenomena. Other conjectures are also possible to explain the data, such as potentials being generated by a deep lying source and propogating to the surface at a later latency, but it appears that two or more generators are responsible for the evoked potentials; for in the spatial ensemble correlation matrices we have observed that significantly correlated groups of potential arise and subside. These transient, smaller groups of high correlations may be an indication that there is a variety of generators located in the brain which contribute to the evoked potential at different latencies. These short lived, highly correlated islands would not have as great an effect on the VER as would the two major rotations discussed earlier. These weaker sources may be the cause of the minor pertubations that are found between evoked responses for slightly differing stimulus conditions.

In summary it appears that there are two major sources that contribute to the SVER at different latencies. The evidence seems to indicate that potential spread on the scalp is active, and these potentials are postulated to be generated by sources firing at different latencies and/or at different locations.

MODELING

It is difficult to construct models of cortical activity which reflect the functioning and interaction of neuronal ensembles at a physiological level because so little is known. However, modeling is more feasible with potential field data since one model is more likely to handle all normal subjects since this data has less inter-individual variability than single evoked potentials.

Two basic types of models have been postulated to explain the rotating characteristics of the data. The first is a gross dipole model, where either a rotating dipole or a combination of fixed dipoles and a quadrapole acting as sources, firing with different delay times or at different depths, could be used to explain the experimental data.

The second model relies nore heavily on anatomical characteristics of the visual cortex. Here the experimental data is modeled by a set of simple sinusoids having different phase shifts. It can be further shown how a model consisting of overlapping potentials generated by EPSP's and IPSP's can account for these phase shifts.

RATIONAL FOR TRANSCRANIAL STIMULATION

The physiological model briefly described above can be tested via transcranial stimulation by a scalp electrode array similar to that used to monitor the evoked electrophysiological data described above. The spatio-temporal pattern evoked by a specific stimulus would be used to control the stimulation array currents. This proposal would differ from that previously used by Brindley and Lewin (1968) and Delgado (1969), in that a spatio-temporal pattern of stimulation would be used rather than stimulating in sequential fashion one array electrode at a time. The specific points of the rational are as follows:

1) Neuronal Ensembles. The variability between individual responses recorded transcranially is conjectured to be due to structural, functional, and experiential differences, unique to each subject. However, the monitoring of the evoked spatio-temporal pattern should illuminate those characteristics of the evoked response common to the population of subjects. It appears that the stimulus parameters are encoded within the response in both a temporal and spatial manner, thus, one problem is to interpret the coding, i.e., to determine the specific locations of ensembles of neurons excited by a particular stimulus and also to monitor the temporal variation of the ensemble. The functional characteristics of the brain can be modeled from data obtained by passive monitoring of ensemble activity. These models must then be tested and stimulation of neuronal ensembles by an electrode array is such a procedure. One might thus be able to test whether the magnitude of an evoked response could be attributed to a larger number of neuronal ensembles being activated, or to larger post-synaptic potentials evoked within the neuronal ensemble, to a higher degree of synchronization of neuronal ensemble firing.

2) Stimulation and Perception. This procedure would more rapidly establish perceptual correlates reflected in electrophysiological measurements with the stimulation pattern.

3) The Electrode Configuration. It has recently been well established by several investigators that there is a direct comparison between cortical surface and scalp potentials. The past literature has been controversial in this area, however, it now appears that when "large" electrodes are used to record the potentials the potentials appear the same regardless of the placement on the scalp or cortical surface. If needle-like or microelectrodes are used on the cortical surface and a large electrode is used on the scalp then the two potentials are dissimilar. However, if one increases the number of small electrodes used on the cortical surface and ties their leads to a common amplifier, thus, effectively increasing the area of the cortical surface monitored, then both cortical and scalp potentials become more similar. Thus, from this and other data

obtained from animal studies, it appears that the skull and scalp do not distort cortical potentials measured from the scalp but merely attenuate them.

4) Functionally Intact Human Subject. The subject would experience no surgical insult, be under no anesthesia, and would be fully functional.

Problem areas do exist and these would include electrical insults to the subject due to stimulation and possible side effects, however, it is the purpose of the Neuroelectric Society and this conference in particular to provide information in this area. If such stimulation proves successful then the door would be open to develop a prosthetics device to aid the blind.

REFERENCES

Bourne, J.R., Spatio Temporal Characteristics of the Visual Evoked Response in Man, Doctoral Dissertation, University of Florida, 1969.

Brindley, G.S., and Lewin, W.S., The Sensations Produced by Electrical Stimulation of the Visual Cortex. J. Physiol., 1968, 196: 479-493.

Clynes, M., and Kohn, M., Spatial Visual Evoked Potentials as Physiological Language Elements for Color and Field Structure, In W. Cobb and C. Morocutti (eds.), The Evoked Potentials (Electroenceph. clin. Neurophysiol, Suppl. 26), Elsevier, Amsterdan, 1967; 82-96.

Delgado, J.M.R., Radio Stimulation of the Brain in Primates and Man, Anesthesia and Analgesia, 1969, 48: 529-542.

Gastaut, H., and Regis, H., Visually-evoked Potentials Recorded Transcranially in Man. In L.D. Proctor and W. R. Adey (eds.), The Analysis of Central Nervous System and Cardiovascular Data Using Computer Methods (1964 Symposium). NASA, Washington, D.C., 1965: 7-34.

Remond, A., Topological Aspects of the Organization, Processing and Presentation of Data. In L.D. Proctor and W.R. Adey (eds), The Analysis of Central Nervous System and Cardiovascular Data Using Computer Methods (1964 Symposium). NASA, Washington, D.C., 1965: 73-93

IMPEDANCE MEASUREMENTS OF TISSUE

Robert Cohn and Harrold S. Leader

U.S. Naval Hospital, Bethesda, Maryland and
U.S.P.H.S., National Center for Chronic Disease
Control, Arlington, Virginia

Impedance properties of tissue have been fairly extensively investigated (1,2,3). In general it has been difficult
to obtain consistent data at very low frequencies of oscillation;
and all measurements have had to be technically sophisticated
and time consuming, because of the null methods used. These
fundamental problems have limited the routine use of tissue
impedance measurement. The introduction of the Hewlett-Packard
Vector Impedance Meters have made tissue impedance measurements
consistent, reproducible and rapid. A functional schematic of
this instrument is shown in figure 1. The frequency range is
continuous between 5 and 500 kHz, the impedance magnitude range
is 1 to 10 megohms and the impedances and phase angles are

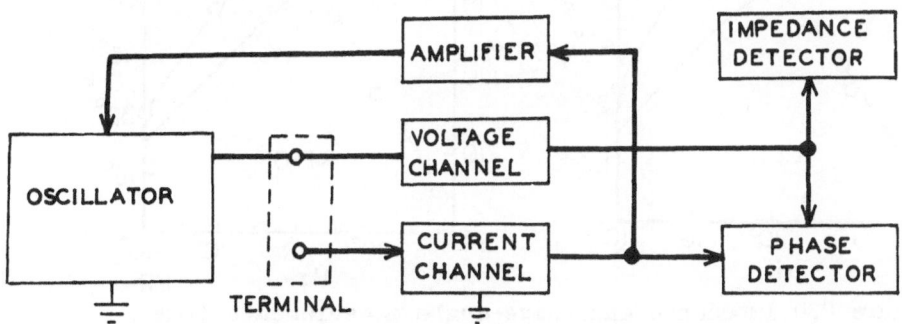

Figure 1. Functional schematic. Constant current
below 1000 ohms; constant voltage above 1000 ohms.

read directly from large meters on the face of the instrument. No specific nulling is employed in these measurements, and a significant set of Z and θ values can be obtained and repeated (for checking) in minutes. The rapidity of measurement allowed us to accumulate an adequate number of data points without undue fatigue to the patient under study.

Using the Vector Impedance Meter 4800A we have determined the impedances between scalp and other body surface electrodes (of various sizes) in more than 40 subjects. Most measurements were made from scalp derivations using 7 mm solder electrodes which were applied to the head with an interposed conducting material consisting of EKG electrode jelly, bentonite and a sufficient amount of concentrated sodium chloride solution to give the material a loose paste consistency.

With electrode spacing ranging from 7.5 to 24 cm between centers, transverse (interhemispheric) and/or rostral caudal (intrahemispheric) derivations showed impedance values of similar orders of magnitude.

The general pattern of impedance change with frequency (in ohms) is shown in figure 2. At 5 Hz, Z ordinarily measured in the order of 50,000 ohms; there was a non-linear decrease in Z with frequency, and in nearly all instances at 500 kHz, Z was in the order of 150 to 300 ohms. This non-linear impedance function was generally similar when log-log plotted from homologous regions of the head. At times there was a two-fold difference in magnitude from symmetric derivation points. In linear plotting

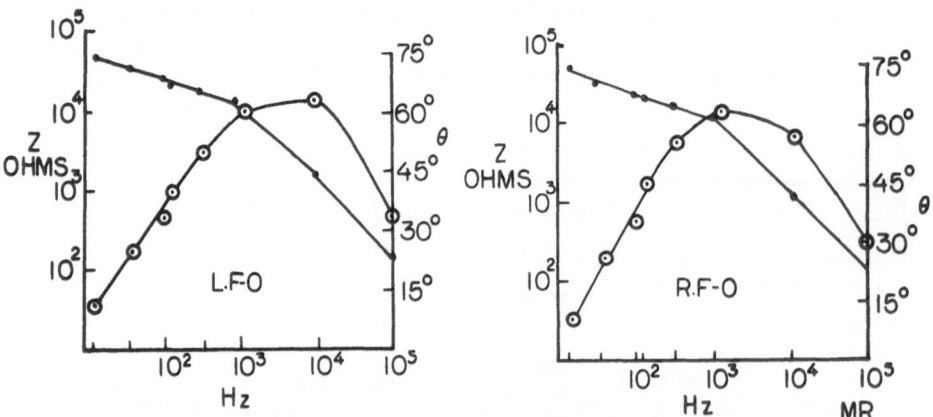

Figure 2. Impedance and phase angle measurements from the right and left frontal-occipital derivations. θ is represented by circled points. Electrodes separated by approximately 25 cm.

the Z function showed a hyperbolic form. The phase angle characteristic was quite similar in all tissues measured; it was
almost invariably negative. This function is shown in the
figure by the circled points. In nearly all cases there was
a maximum at 1 kHz to 10 kHz, with a rapid decrease of θ at
100 kHz.

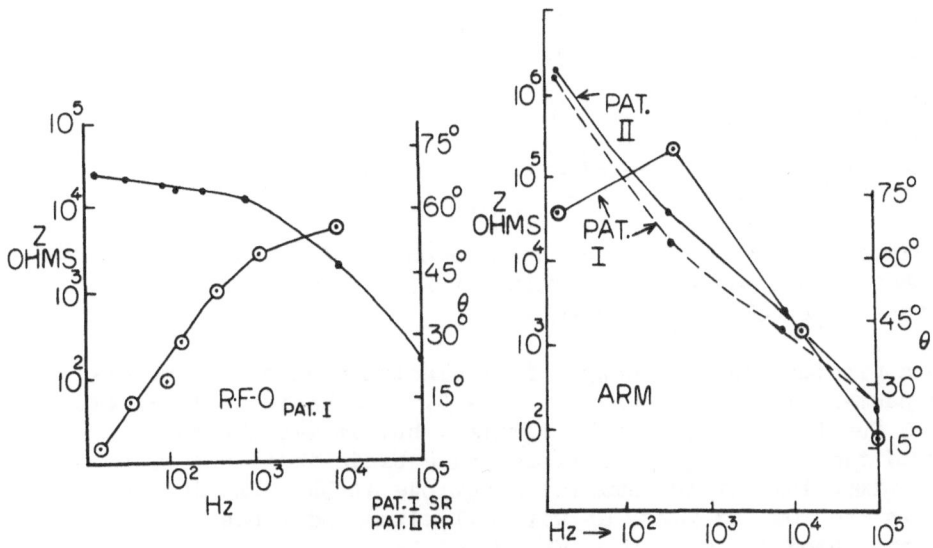

Figure 3. Z and θ for scalp derivation (left side of
figure) and arm derivations (right side of figure).
Note the log order of magnitude difference of Z at 5Hz.

Figure 3 shows the Z and θ measurements from head derivations (left part of figure), compared with the arm derivations
(right part of figure). In each instance the electrodes were
separated by equal distances. The absolute value of Z from
the scalp at 5 Hz was around 40K ohms; the arm absolute value
measured around 10 megohms in each patient of the figure. In
all cases where the scalp and arm surface recordings were made
in the same subject the arm had an approximately 10 to 20-fold
greater absolute impedance value at 5Hz. At 100 kHz, the Z
values were always of a similar order of magnitude at around
100 to 300 ohms. The θ function of the arm derivations showed
the usual rise to maximum followed by a decreased negative angle
with increased frequency but was less smooth than θ from
corresponding scalp electrodes.

The effect of the pick-up area of the surface electrode
on Z and θ was measured using the ordinary, approximately 7 mm
diameter solder disk with underlying paste, and an approximately

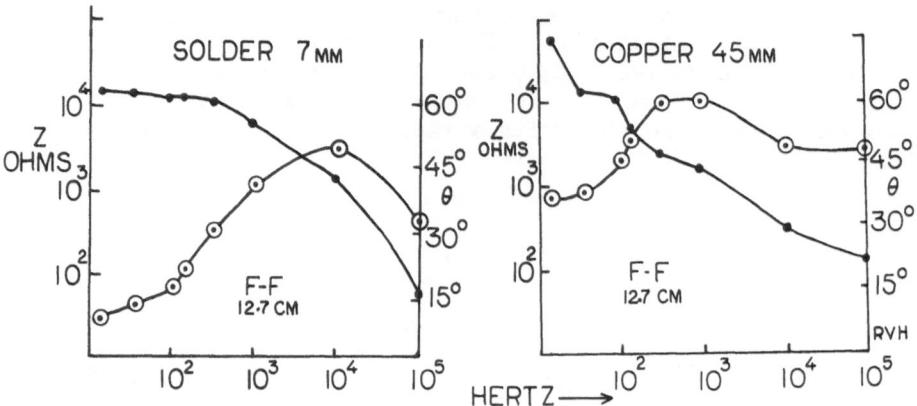

Figure 4. Effect of pick-up area on Z and θ. Note
that at 5Hz the absolute value of Z is greater for
the larger electrode surface.

7 times greater diameter copper foil electrode also with underlying paste. This is shown in figure 4. The basic contours of
Z and θ functions were grossly similar, but it was always
observed that the larger electrode had a basic impedance of
3 to 4 times that of the smaller electrode at 5Hz. At 100 kHz,
Z was in the 100 ohm range as with all sizes and types of
electrodes used.

     Because of the use of pointed metallic electrodes in many
physiological studies, measurements were made of 150 micra
diameter stainless steel wires rapidly tapered to around 1
micron points. Such electrodes are routinely used in our
laboratory. When the pointed uncoated needles were placed in
normal saline and separated by 25 mm the Z and θ curves of the
left side of the figure 5 were obtained. It is observed that
θ recedes with frequency, just as does Z. When the needles were
insulated to the tip with baked varnish θ maintained an almost
constant value, whereas Z, which had an initial higher value
than the uncoated electrode, showed a non-linear decremental
character.

     We did not routinely extend the measured frequency beyond
100 kHz in this work, but occasionally we did use the 500 kHz
limit. Under these latter conditions Z continued to decrease
in value. It appears of some interest to subsequently determine
Z and θ for much higher frequencies of oscillations. The
possibility however arises that with much higher frequencies
in the gigacycle range, the dermal penetration would become
negligible, and as a consequence the tissue impedance measure-

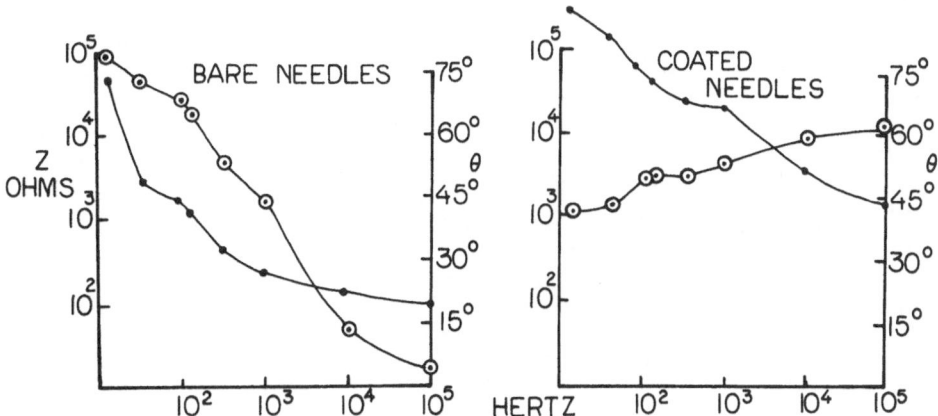

Figure 5. Impedance and phase angles of coated and
uncoated pointed stainless steel were in saline.
Electrode separation was 25 mm.

ments, might be quite spurious. If indeed the tissue impedance
did approach zero with high frequency stimulation it appears
feasible that the low voltage spontaneous and evoked biological
potentials might under certain non-linear conditions modulate
an impinging coherent carrier frequency. Under these conditions
the biologically generated potentials might be demodulated
and recorded from a distance without direct electrode conduction.

In summary the impedance measurements of all skin
derivations show a basically non-linear characteristic change in
frequency. The absolute values of Z varied through a relatively
narrow range from patient to patient, however the extremity
skin always showed a higher value by an order of 50 to 100 times
that from the scalp derivations.

The ease, consistency and reliability of the Z and θ data
obtained would appear to merit more extensive use in laboratories
devoted to neuro-physiology and its related disciplines.

REFERENCES

1. Burns, R. C. "A Study of Skin Impedance". Electronics 23:
    190-196, 1950.
2. Cole, K. S. and Curtis, H. J. "Electrical Physiology:
    Electrical Resistance and Impedance of Cells and Tissues."
    Medical Physics. Vol. 1. The Year Book Publishers, 1944.
3. Schwan, H. P. "Physical Techniques in Biological Research."
    Academic Press, New York, 1961.

ELECTRICAL ACTIVATION OF SKELETAL MUSCLE BY SEQUENTIAL STIMULATION

P. H. Peckham, J. P. Van Der Meulen, and J. B. Reswick

Case Western Reserve University

Cleveland, Ohio, U. S. A.

INTRODUCTION

The ability to voluntarily contract skeletal muscle is lost
when the neural connection between the brain and muscle is impaired.
Functional Motoric Electrical Stimulation of the Extremities has been
proposed as a method of inducing controlled movements in paralyzed
limbs by the application of electrical current to the muscle (1).
However, a major obstacle in applying such a method clinically is
that the muscle rapidly loses its contractile strength. Although
the causes and mechanisms of fatigue are not well understood, the
repetitive firing of muscle fibers at their fusion frequency and
the occlusion of blood flow are likely responsible (2). Sequential
stimulation, a technique of electrical activation of skeletal muscle
in which the stimulus is multiplexed through a number of electrodes
within the body of the muscle, is designed to minimize these effects.
We have performed experiments in animals to determine the feasibil-
ity of sequential stimulation and preliminary experiments in humans
to evaluate an electrode system for chronic stimulation.

MODULATION OF THE STRENGTH OF AN ELECTRICALLY INDUCED CONTRACTION

The unidirectional, regulated current, rectangular waveform is
widely used both clinically and experimentally. To completely describe this waveform it is necessary to specify the pulse amplitude,
pulse duration or width, and frequency or repetition rate. The
strength of the contraction can be varied by changing any of these
parameters. These methods are called amplitude modulation, pulse
width modulation, and frequency modulation respectively.

If a single active electrode is applied to a particular muscle,
the frequency of the stimulus to obtain a fused force output of the

45

muscle has some minimum. Increases in frequency beyond the fusion frequency yield only small increases in the strength of contraction. Thus, frequency modulation is not a useful technique for monopolar stimulation, and frequency is usually fixed at the fusion frequency.

Pulse width modulation also has a limited range of effectiveness. The magnitude of the increase of the strength of contraction gradually decreases for incremental increases in the pulse width when the pulse amplitude is held constant(3). Initial work has shown that more power must be dissipated and charge transferred through the tissue to obtain the same strength of contraction for pulses either longer or shorter than 0.1 msec. These factors have been implicated to produce tissue damage which should be minimized for chronic stimulation. The pulse width is then fixed at 0.1 msec.

The remaining monopolar technique for controlling the strength of contraction is amplitude modulation. The rate of fatigue is rapid with this method because the blood supply to the muscle is occluded with the stimulus frequency necessary to obtain a fused contraction (2). Further, large stimulus amplitudes will result in the spread of the stimulus to neighboring muscles.

The insertion of a number of electrodes into the body of the muscle yields an additional technique known as sequential stimulation. Figure 1 shows the mechanism that is likely involved in sequential stimulation. This is not actual data, but is indicative of what is experimentally observed. The concept of sequential stimulation is to cause fibers of the muscle to contract less often than is required by stimulation through a single electrode. By inserting a number of electrodes in the muscle, in this case three, and having each fire a different portion of the muscle by a sequential switching of the stimulus through the electrodes, a fused response can be obtained from the subtetanic and unfused responses of each individual portion of the muscle. In addition, if the portions of the muscle that are activiated by each electrode are non-overlapping, since each portion is fired less often, the muscle fatigues less rapidly than it does with single electrode stimulation.

In general, the contractile response of a muscle to a stimulus applied at two different points in the muscle is not the same. Thus, to obtain a smooth fused contractile response it is necessary to adjust the stimulus parameters for each electrode in the muscle so that all partitions of the muscle produce approximately the same strength of contraction. Then the total response of the sequentially stimulated muscle will be a smooth, fused contraction.

METHOD

The hind limb of anesthetized cats (Sodium Pentobarbital, 40 mgm/kg intraperitoneal) was held in the horizontal plane by two clamps gripping the bony prominences of the lower leg. The tendons of all muscles acting across the ankle except the tibialis anticus

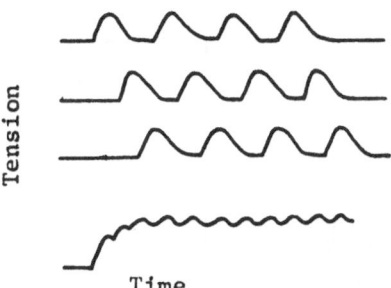

Figure 1.   Summation of Tensions Developed by Three Electrodes
            Firing Sequentially.

muscle were severed. The angle of the ankle was fixed at 120 degrees (approximately the neutral position) by a strain gauge bar
firmly connected to the foot. Nearly isometric contractions of the
tibialis anticus were recorded by a type Q Tektronix AC bridge amplifier as a torque about the ankle. Minihelix coil stainless steel
(type 304) cathodic electrodes 25 mm in length with a deinsulated
portion of 10 mm and diameter of 0.152 mm were inserted into the
tibialis anticus and a 6 cm indifferent anode of silver wire covered
with saline soaked felt was placed on the skin surface over the
thigh muscles (4). The coil electrodes were spaced equal distances
of approximately one centimeter apart at different depths of the
muscle. The pulse duration was set at 0.1 msec and the amplitudes
of the stimuli applied to each electrode were adjusted until the
contraction strengths were the same for each portion of the muscle.

    The fatigue induced by sequential stimulation was compared to
that induced by stimulation through a single stimulating electrode.
The peroneal nerve was isolated by blunt dissection and two electrodes were applied along its length. A supramaximal stimulus was
delivered to the nerve to determine the maximal contractile strength
of the tibialis anticus. With a stimulus frequency of 60 Hz, the
stimulus amplitude applied to each electrode was then adjusted until the contraction strength was 25 percent to the peak contractile
strength. The stimulus frequency applied to each of the four electrodes during sequential stimulation was then set at 15 Hz. One
second bursts of stimuli were alternated with one second rest periods
for a total time of two minutes. Four muscles were fatigued with
the single electrode technique and three using sequential stimulation. The results were averaged.

RESULTS AND CONCLUSIONS

    The frequency of the stimulus applied to each electrode was
varied from 5 to 80 Hz. Figure 2 shows the dependence of contraction strength on the frequency of the stimulus for two sets of
stimuli applied to four electrodes. Ia, Ib, Ic, Id are the amplitudes
of each of the stimuli. A fused contraction was obtained with a

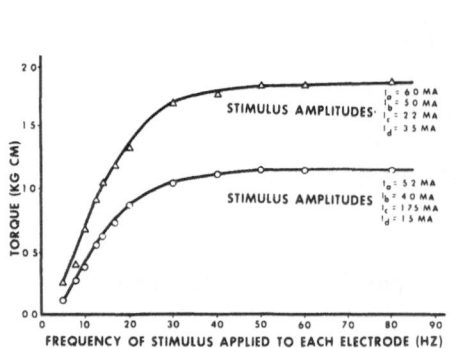

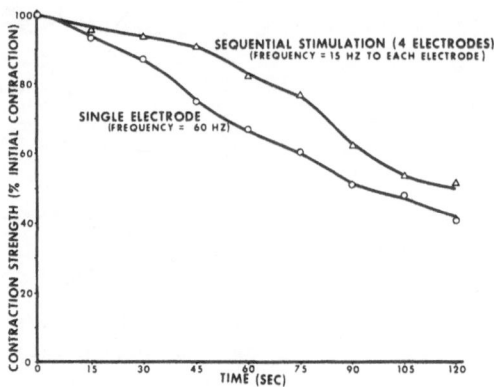

Figure 2.   Frequency Modulation of   Figure 3. Comparison of Fatigue
Contraction Strength Using Sequen-   Induced by Intermittent Stimu-
tial Stimulation (4 Electrodes)   lation for 25% Maximal Contraction

stimulus frequency of 10 Hz applied to each electrode, which is
equivalent to a frequency of 40 HZ (10x4) applied to the entire
muscle. The slope of the curves is zero for frequencies above 50
or 60 Hz, which is the tentanic frequency for the tibialis anticus
stimulated with a single active electrode. For this muscle, the
strength of contraction can be modulated using frequencies from a
minimum of 40 Hz applied for the entire muscle to 60 Hz applied to
each electrode.

     Figure 3 shows a comparison of fatigue induced by intermittent
stimulation with a single electrode and sequential stimulation for
an initial contraction twenty-five percent of the maximal strength.
The muscle fatigues less rapidly during sequential stimulation than
during stimulation with a single electrode. The difference between
the end points of the curves is only ten percent. However, one
could expect significant improvement when the electrodes are carefully positioned in a larger muscle such as that of a human.

TISSUE REACTION STUDIES

     The effect of chronic implantation of the helical coil elec-
trodes within the muscle was histologically examined. Excellent
tissue compatibility between the electrode and the tissue was shown
both with and without the passage of an electrical current within
the functional range.

A PERCUTANEOUS ELECTRODE SYSTEM FOR CHRONIC STIMULATION

     A percutaneous electrode system was implanted in the forearm
of one normal human subject to evaluate direct stimulation of muscle.
This system consists of three stainless steel helical coil electrodes which passed through the skin into the extensor digitorum

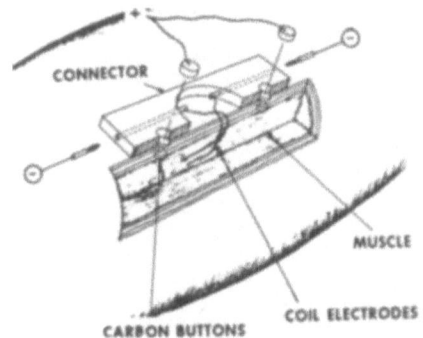 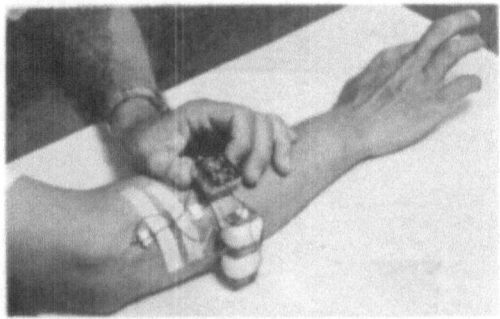

Figure 4. Percutaneous Electrode System for Chronic Stimulation

communis muscle, and two vitreous carbon "buttons" which formed a common anode and firmly anchored the surface connector. Passage of current resulted in an extension of the middle and ring fingers without the discomfort associated with stimulation through the surface of the skin (Figure 4). This system remained intact within the subject's forearm for 18 weeks, at which time the experiment was terminated. Results, inconclusive at this time, indicate that sequential stimulation markedly reduced the fatigue. This electrode system is believed to be the first chronic electrode arrangement for direct skeletal muscle stimulation in humans.

This work was supported by Grant Number RD-1814-M from the Social and Rehabilitation Service, Department of Health, Education, and Welfare.

1. Liberson, W. T., Holmquest, H. J., Scot, D., Dow, M., "Functional Electro Therapy: Stimulation of the Peroneal Nerve Synchronized With the Swing Phase of the Gait of Hemiplegic Patients," Archives of Physical Medicine and Rehabilitation, Feb. 1961, pp. 101-105.

2. Kugelberg, E., and Edstrom, L., "Differential Histochemical Effects of Muscle Contractions on Phosphorylase and Glyogen in Various Types of Fibers: Relation to Fatigue," J. Neurol. Neurosurg. Psychiatry, 1968, 31, 415-423.

3. Peckham, P. H., "Design Considerations in Electrical Stimulation of Skeletal Muscle," Case Western Reserve University, Report No. EDC-4-68-23, June, 1968.

4. Caldwell, C. W., Reswick, J. B., "A New Transcutaneous Electrode," In M. Eden (Chairman), Proceedings of the 20th Annual Conference on Engineering in Medicine and Biology, Joint Committee on Engineering in Medicine and Biology, Boston, 1967, 9:15.6.

A THEORETICAL ANALYSIS OF SEVERAL ELECTRODE PLACEMENTS ON THE SCALP FOR RECORDING AVERAGED EVOKED POTENTIALS

Daniel A. Driscoll

Union College

Schenectady, N.Y.   12308

In the following discussion, theoretical data will be presented comparing the relative effectiveness of mono-polar and closely spaced bi-polar electrodes in localizing the position of a region of evoked cortical activity. The basis for comparison of the two electrode systems is the ability to reject signals produced by sources of electrical activity other than those cortical sources directly below the recording electrode(s). It is assumed that the maximum sensitivity of the two systems is a relatively unimportant consideration since sensitivity is effectively increased by commonly used averaging techniques.

The theoretical evaluation of the electrode systems is performed with the aid of a three-concentric-spheres mathematical model of the head [1,2] using parameters for a typical human head: head radius-- 9.2 cm; skull thickness--0.5 cm; scalp thickness--0.7 cm; brain and scalp resistivity--222 ohm-cm, and skull resistivity--17,760 ohm-cm. In the model, ideal point electrodes are used for the surface potential measurements, and each small area of cortical activity is represented by a single dipole source, the position of which is varied; at each position, the dipole is oriented for maximum response.

BI-POLAR ELECTRODES

The bi-polar electrodes to be considered have interelectrode spacings of 0.5, 1.0, and 2.0 cm. The sensitivity of the electrode pairs to a dipole source located 0.2 cm below the inner surface of the skull and directly below the midpoint of the two electrodes is 2.4 mV signal strength per V-cm^2 dipole strength for the 0.5 cm electrode spacing; the sensitivity is 4.6 mV/V-cm^2 for the 1.0 cm

51

spacing, and 7.6 mV/V-cm^2 for the 2.0 cm spacing.

The relative response to other dipole sources located along
the line of the electrodes and 0.2 cm below the cortical surface is
plotted in Figure 1.(solid lines). It can be seen from the Figure
that a large improvement in the rejection of peripheral signals is
obtained by reducing the electrode spacing from 2.0 to 1.0 cm; decreasing the spacing again by a factor of two from 1.0 to 0.5 cm,
a spacing less than the 0.7 cm scalp thickness, gives little improvement.

For comparison with other electrode systems it is desirable to
have some figure of merit describing the degree to which peripheral
signals are rejected. From the Figure it can be seen that the sensitivity of the electrodes with 1.0 cm spacing is 50 percent of the
maximum sensitivity if the coritcal source is located 2.0 cm from
the position for maximum sensitivity. The 2.0 cm is an estimate of
the minimum spacing two cortical sources can have and still be recognized as separate sources.

MONO-POLAR ELECTRODES

The relative sensitivity for a mono-polar electrode pair is
also plotted in Figure 1 (dashed line); the position of the cortical
source, located 0.2 cm below the inner surface of the skull, is
varied beginning at the position directly under the recording electrode. The "neutral" electrode is located about 15 cm (90 degrees
along the sphere surface) from the recording electrode and perpen-
dicular to the direction of the source movement. The position of
the neutral electrode can be moved farther from or closer to the
recording electrode without significant change in the shape of the
relative sensitivity curve. The peak of the relative sensitivity
curve for the electrode system described corresponds to a sensitivity
of 12.2 mV/V-cm^2; this peak sensitivity does not decrease substan-
tially until the electrode spacing is less than 5 cm.

The figure of merit for the mono-polar electrode system is 3.3
cm, the distance a dipole source must be located from the position
for maximum sensitivity for a 50 percent decrease in sensitivity.

BI-POLAR VERSUS MONO-POLAR ELECTRODES

As can be seen from the Figure and from the discussion above,
bi-polar electrodes with spacings of 1.0 cm or less have somewhat
greater ability to reject peripheral cortical signals (or to separate closely spaced areas of cortical activity). In addition, the
bi-polar electrode pair rejects thalamic signals by a factor of 15
while the ratio of sensitivities to cortical and thalamic sources
for the mono-polar electrode is only 5:1.

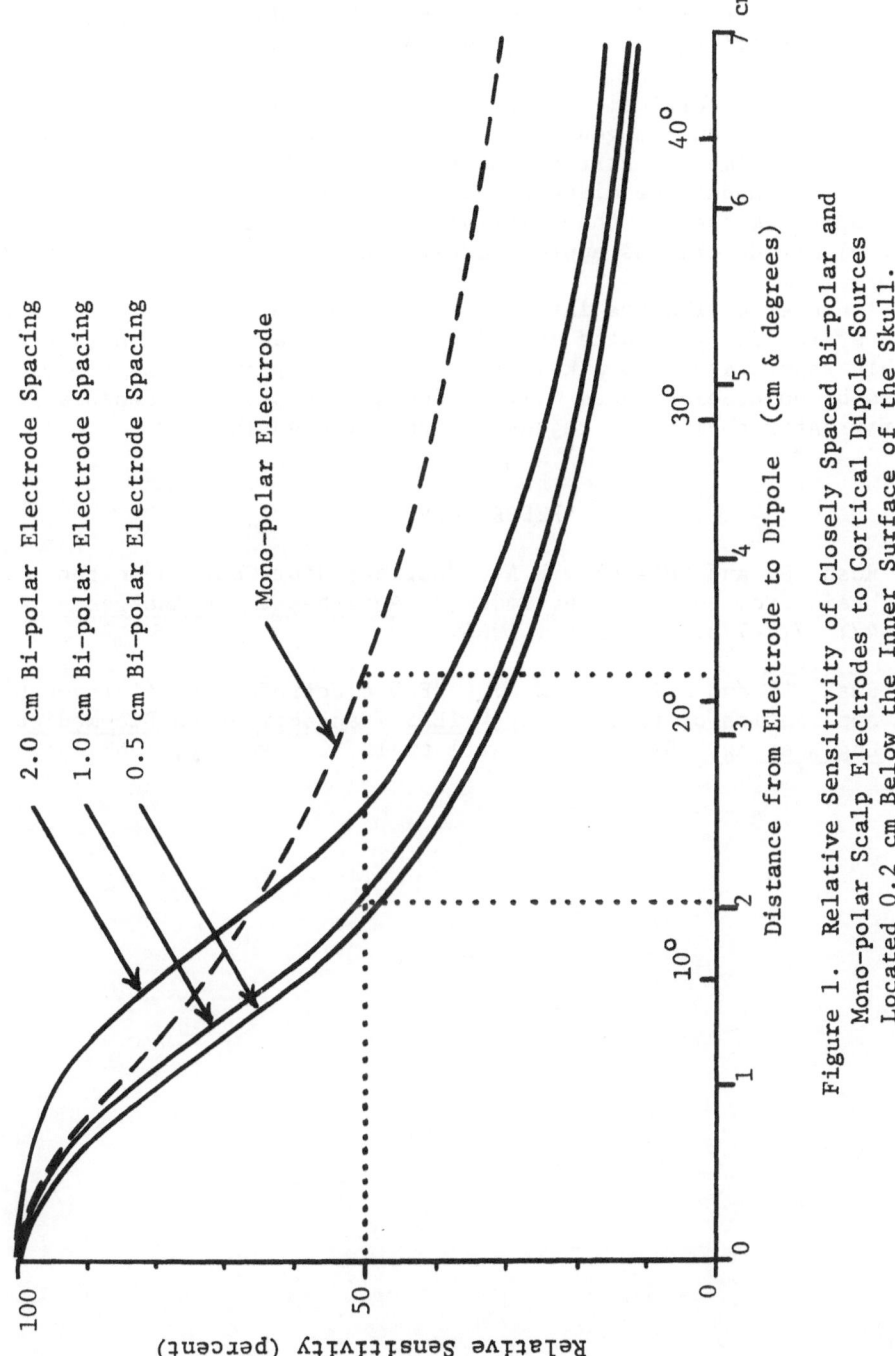

Figure 1. Relative Sensitivity of Closely Spaced Bi-polar and
Mono-polar Scalp Electrodes to Cortical Dipole Sources
Located 0.2 cm Below the Inner Surface of the Skull.

While it is true that the maximum sensitivity of the mono-
polar electrode is about three times the maximum sensitivity of
bi-polar electrodes, it is assumed that the reduction in sensitivity
is not a significant consideration. A consideration which may play
a role in the choice of an electrode system is the fact that at peak
sensitivity the cortical dipole is oriented radially for the mono-
polar electrode and tangentially for the bi-polar electrode. However, by the time the cortical source is moved one centimeter from
the position under the midpoint of the bi-polar electrodes, the
orientation is radial; the orientation of the source for the mono-
polar electrode remains approximately radial.

A number of factors influencing the choice of a suitable electrode system for recording evoked potentials have been considered
in this paper; it seems likely that for many applications the closely
spaced bi-polar electrodes will be found more suitable because of
their greater ability to reject peripheral signals.

REFERENCES

[1] Rush, S. and Driscoll, D.A., "Current distribution in the
     brain from surface electrodes," Anesthesia and Analgesia,
     47: 717-723, Nov.-Dec., 1968.

[2] Rush, S. and Driscoll, D.A., "EEG electrode sensitivity--an
     application of reciprocity," IEEE Transactions on Bio-medical
     Engineering, BME-16, Number 1 pp 15-22, January, 1969.

SECTION 2

NEUROPHYSIOLOGICAL EFFECTS OF ELECTRICAL CURRENTS

EFFECT OF MICROWAVE RADIATION ON THE FROG SCIATIC NERVE

Jeffrey Rothmeier, Ph.D.

Manager, Biomedical Engineering Program

Sanders Associates, Inc., 95 Canal Street, Nashua, N.H.

INTRODUCTION

A variety of biologic effects of microwave energy have been reported. Lens opacities can be induced with 12.3 cm microwaves at power levels of 250 mw/cm^2 (Williams). Alterations in cardiopulmonary, thyroid, and erythropoietic functions have been observed using power levels of 50 mw/cm^2 (Michaelson). Radiation at power levels of 25 mw/cm^2 is sufficient to maintain a 1°C increase in body temperature (Ely).

Effects on the nervous system are very diffuse and do not necessarily increase with increasing power levels. Data obtained on EEG changes in the rabbit (Gvozdikova) and clinical observations based on case histories of humans exposed to microwaves (Turner) are reproduced in the Appendix. These data are very suggestive but not conclusive. There is a need to collect more definitive information on the effects of low level radiation on the nervous system. The preliminary data presented herein, in which effects on the frog sciatic nerve are explored, is a small step toward this goal.

PRELIMINARY RESULTS WITH PULSED RADIATION

This work began with the observation that it is possible to perceive what appears to be an auditory tone when one is placed near certain radars (Ingalls). Two of them with the following characteristics have been heard:

1. Freq = 1.2568 gHz, Peak Power 500 kw Ave. Power 15 mw/cm^2, Dur = 2.5 μsec, PRF = 400 Hz

2. Freq. = 2.8 gHz, Peak Power 200 kw
    Dur = 3-4 μsec, PRF = 1 Hz

The second radar was a model SCR-584 modified to deliver a pulse
synchronously with a pulse generator - usually at the rate of once
per second. A "click" was heard each time a microwave pulse was
emitted. No auditory sensation was obtained from an x-band continuous wave source at an average power of 1.6 mw/cm^2.

It is possible that the sensation above is due to direct
neural stimulation. An attempt to determine whether either of the
two pulsed radar systems above would evoke an action potential in
the frog sciatic nerve was unsuccessful. The twitch of the gastroc-
nemius muscle was used as an indicator in this case.

An experiment to determine whether or not the 2.8 gHz pulsed
radar affects the excitability of the sciatic nerve was next
attempted. The nerve was placed in a plexiglass nerve chamber
about one half meter from a 10 cm. open end waveguide. Table I
indicates that the radiation increased the excitability of the nerve.
These results are not conclusive because it was not possible to
monitor the electrical test pulse which occurred concommitantly
with the microwave pulse, the wires which carried the electrical
test pulse were exposed to the microwave field, and precise measurements of the power of the microwave field were not made. Monitoring
the electrical test pulse is important because the SCR-584 radar
produced a considerable transient whenever a pulse was emitted.

TABLE I

Effect of Microwave Stimulation on the Strength-Duration
Curve

| Stimulation Strength (mw) | Duration (μsec) | |
	No Radiation Present	Radiation Present
2000	10	10
800	12	11
280	42	39
80	170	135
28	---	620
9	---	---

Another experiment in which subthreshold electrical pulses
were combined with the microwave pulses is illustrated in Figure 1.
The microwave and electrical pulse combination did not always induce
an action potential. One minute of 1/sec stimulation was sometimes
required before the response in Figure 1 was observed. Further,

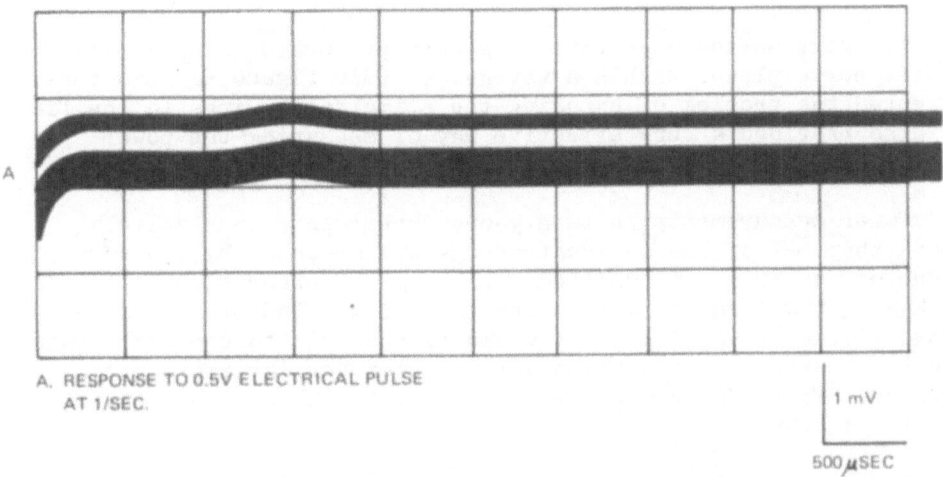

A. RESPONSE TO 0.5V ELECTRICAL PULSE
   AT 1/SEC.

1 mV

500 μSEC

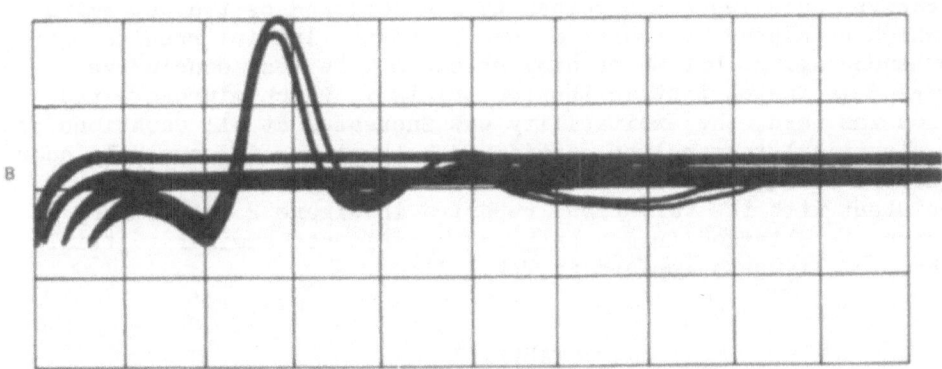

B. RESPONSE TO 0.5V ELECTRICAL PULSE
   AND RADAR PULSE AT 1/SEC.

Figure 1.   Effect of Radar on the Response of the Nerve
            to a Subthreshold Electrical Stimulation

once an action potential was obtained, it did not occur at every
stimulation.

USE OF CONTINUOUS WAVE RADIATION

The next series of experiments were performed using a frog
sciatic nerve placed within a waveguide. (see Figure 4) This cir-
cumvented the problem of exposing the stimulating wires to the field
and also provided a very effective way of measuring the power
absorbed by the nerve.

Power measurements in twenty-one experiments conclusively
showed that 52% of the incident energy was absorbed by the nerve.
If one of the trials is omitted, the range of absorptions obtained
is 46% to 57%. In all cases, the total power injected into the
x-band waveguide was 1 mw. It would be possible to use this system
to obtain curves of absorption versus thickness for small cross
sections. These could then be compared with the semitheoretical
results of Schwan.

The instability of the excitability of the frog nerve with time
causes some problems in assessing the effects of microwave stimulation.
(Figure 2) The determination of the strength duration curves with
microwave radiation was preceded by and followed by two determinations
in which no microwave radiation was present. Initial results were
very encouraging, but succeeding attempts were less conclusive.
Figure 3 indicates that in the two trials in which microwave radiation was used, the excitability was increased at all durations of
the electrical test pulses considered. About one and one half hours
transpired between each trial - the variability in the data is
consistent with the variations reported in Figure 2. A summary of
the results obtained from twenty-one experiments similar to trials
2 and 3 of Figure 3 appears in Table II.

TABLE II

Effect of .43 cm/cm^2, 10 gHz, Continuous Wave Radiation on
the Excitability of the Frog Sciatic Nerve

A.   Trials in which the threshold was set at the minimum
detectable response

Possible increase in excitability   3
No change in excitability   2

B.   Trials in which the threshold was set at an action potential
amplitude of .4 mv

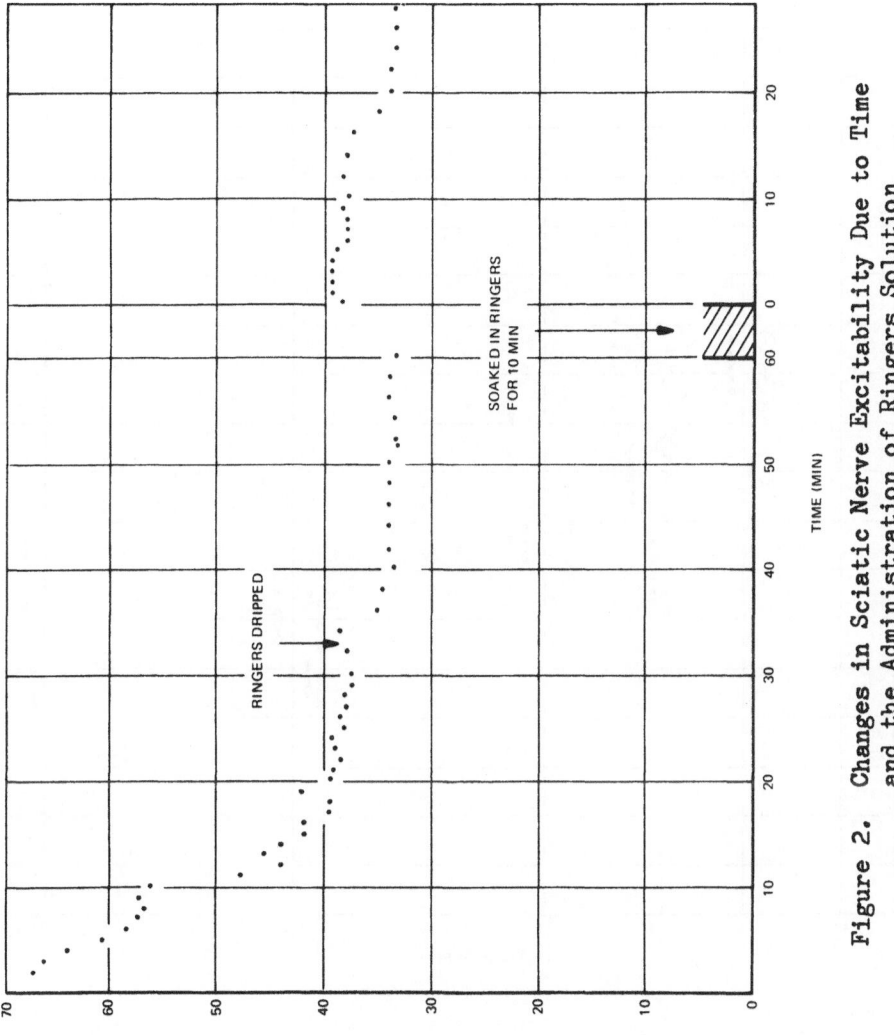

Figure 2. Changes in Sciatic Nerve Excitability Due to Time
and the Administration of Ringers Solution

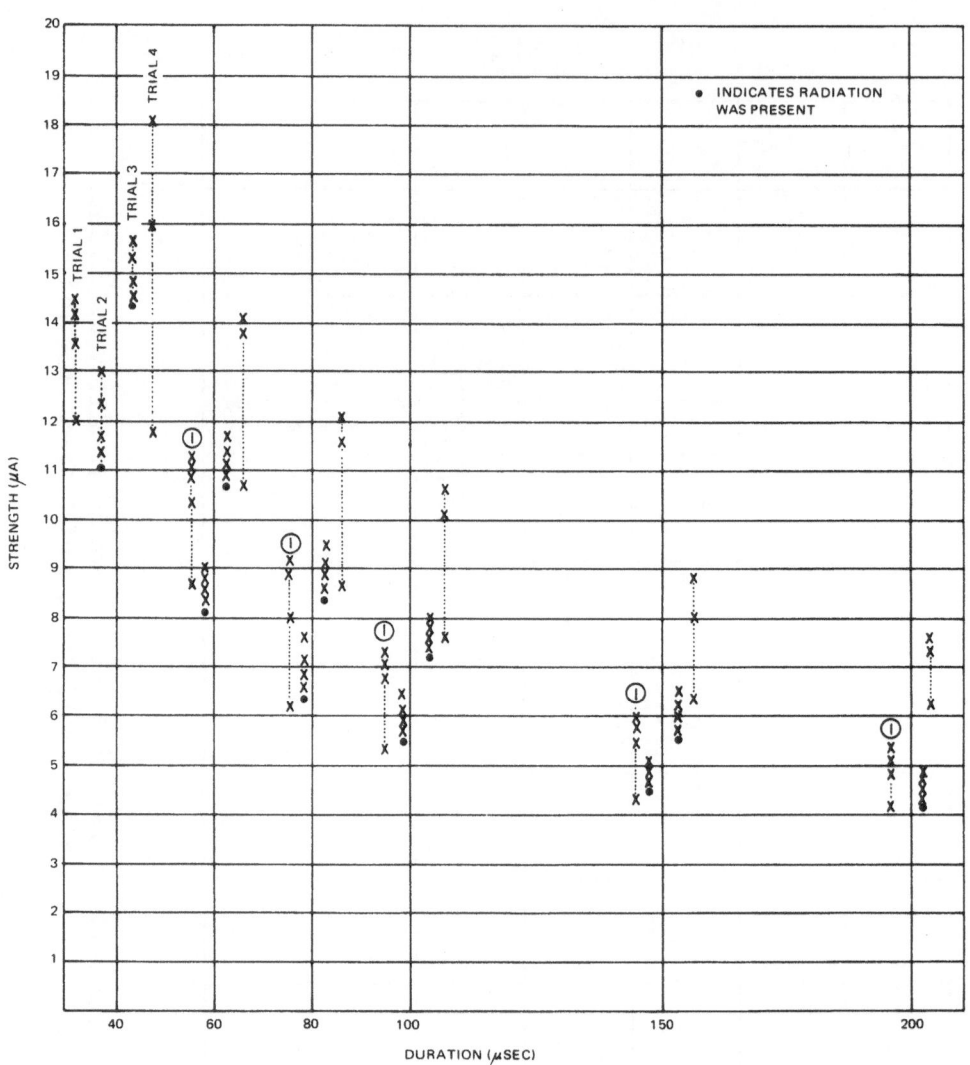

Figure 3.   Effect of .43mw/cm^2 Continuous Wave, 10 gHz
Radiation on the Excitability of the Frog
Sciatic Nerve

TABLE II (cont'd)

Possible increase in excitability	6
No change in excitability	10

The trials in which increases were reported were not in general as conclusive as those illustrated in Figure 3. Data from two typical trials are provided in Table III. Radiation is present only during the determination of the third column of strength data–a period of about ten minutes.

TABLE III

A. Trial in which a change in excitability due to microwave radiation may exist.

Duration (μsec)	Strength (μA)				
	1	2	3	4	5
40	1.14	1.22	1.16	1.12	1.13
60	.85	.88	.85	.80	.81
80	.69	.72	.66	.63	.66
100	.59	.60	.57	.55	.57
150	.45	.47	.43	.42	.45
200	.39	.40	.36	.36	.39

B. Trial in which no change in excitability exists

Duration (μsec)	Strength (μA)				
	1	2	3	4	5
40	.96	1.01	1.12	1.16	1.16
60	.70	.73	.79	.80	.82
80	.56	.59	.63	.64	.65
100	.47	.50	.53	.54	.57
150	.37	.40	.41	.43	.44
200	.32	.33	.35	.37	.37

Part A of Table III indicates that if the decrease in excitability in column three for durations 80, 100, 150, 200 is due to the radiation and not normal time fluctuations, then the radiation effects persist beyond the period of radiation (column 4). It is considerations of factors such as residual effects of radiation, technical difficulties of reporting results read from an oscilloscope due to the low signal to noise ratio, and the considerable normal fluctuations of the excitability with time that render these results inconclusive. Other variables, such as the length of time between the experiment and removal of the nerve from the frog and the amount of Ringers solution on the nerve may also be important.

DISCUSSION AND FUTURE PLANS

Kamenskiy has reported the following results using a similar preparation:

"Research on irradiation with continuous microwaves

A. Research on thresholds of excitability.
Radiation was done with microwaves having a wavelength of 12.5 cm and an intensity of 11 mvt/cm^2 for 20 minutes. No changes in the threshold of excitability which would exceed the degree of variation of the threshold in the norm, were discovered in 34 experiments."

"Studies of irradiation with pulse microwaves.

A. Research on thresholds of excitability.
Irradiation was carried out for 20 minutes with microwaves of λ = 10 cm, at a pulse length of 1 microsecond and frequencies of 100, 200, and 700 pulses/sec. The average irradiation intensity was 12 mvt/cm^2.

A certain increase in excitability (15%) was detected when irradiated at 700 pulses/sec. This change is shown in the graph (this graph did not appear in the translation) in which one can also see a tendency for a return of the threshold values of the initial level after stopping irradiation."

The results of pulsed radiation are in agreement with those presented here. It is, however, peculiar that no effects are reported for pulse repetition rates of 100 and 200 Hz. One also wonders whether the average power was increased when the PRF was increased from 100 to 700 Hz, or whether the peak power, which was not specified, was decreased to maintain a constant average power.

To help overcome the effects of time instabilities and increase the accuracy of data taking, an experiment has been devised which utilizes a PDP-9 to collect and statistically analyze data. In this experiment, the amplitude of the action potential before and after radiations will be observed. The instruments used in the experiment are illustrated in Figure 4. With the exception of the computer, it is the same as that used in the experiments with continuous wave radiation reported above.

Average response computing to increase the signal-to-noise ratio and increased speed of collecting data are the greatest benefits achieved from using the computer. The results of this experiment will provide a definitive answer to some of the important parameters for microwave stimulation of the nervous system.

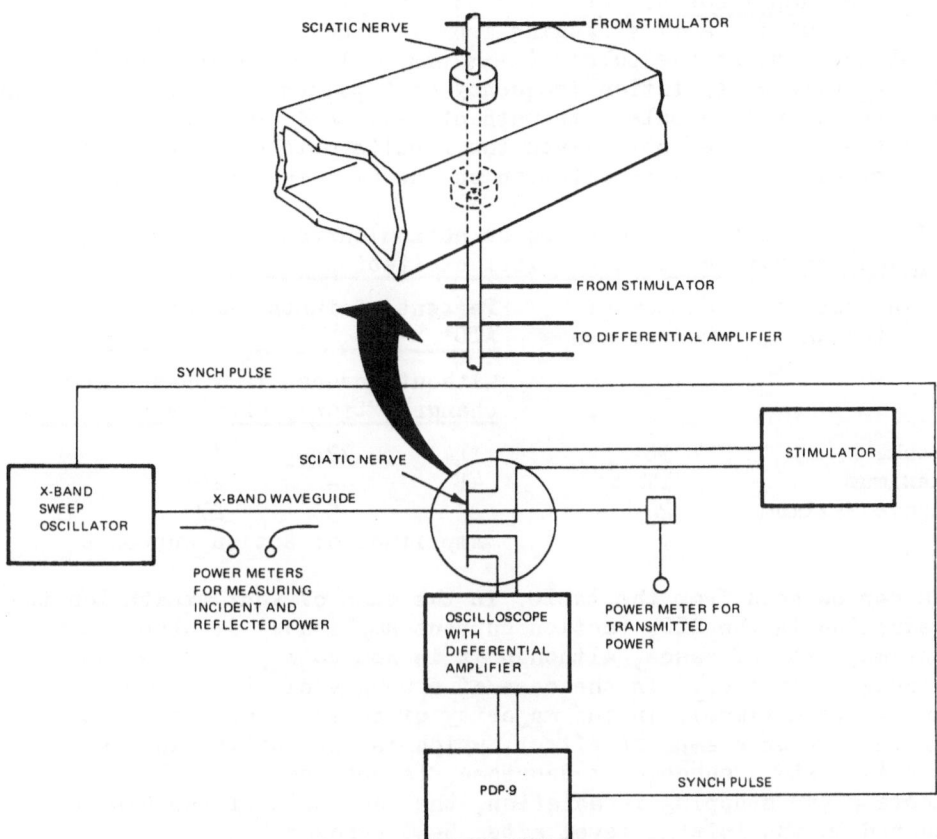

Figure 4.   Instrumentation Configuration

In order to effectively test excitability with a computer, it would be necessary to use it to control the stimulator - hence, this experiment was delayed to a future date. In addition, Kamenskiy has reported that the action potential amplitude is affected by low level continuous wave radiation. His results are quoted below:

"Study of amplitude of nerve action currents. The irradiation was carried out for a time ranging from 10 seconds to 1 minute with continuous microwave pulses (wavelength 12.5 cm) lasting from 1 msec with a repetition frequency of 5 pulses/sec. The radiation intensity, with a pulse strength of 14 vt/cm^2, was 70 mvt/cm^2 on the average. The table gives the results obtained regarding the effect of microwave irradiation on the amplitude of action currents.

Effect of microwave radiation on action current amplitude with various magnitudes of electrical excitation

Magnitude of excitation	Number of measurements	Percent of instances of changes in ATD*			
		without change	reduction	two-phase changes	increase
Weak	36	11	72	17	--
Maximum	17	41	--	24	35
Above Maximum	22	27	5	27	41

*Amplitude of action currents

As can be seen from the table, in the case of weak excitation the reduction in the ATD, (action current amplitude) is observed in the majority of cases, although it is not noted in the event of strong excitation. In the case of strong excitation, (maximum and above maximum), in the majority of cases either an increase in the ATD or a 2-phase effect, which is the initial increase in the ATD with a subsequent decrease, is observed. In all experiments after stopping irradiation, the magnitude of the ATD returned to the initial level after 5-60 seconds."

These results have some peculiar features. Why does radiation affect the action currents obtained from above maximum stimulation greater than those obtained from maximum stimulation? Why are such a large number of instances reported without an affect?

It is always difficult to precisely compare results from two sources, because it is unusual for two people to do things precisely the same way. In the area of microwave effects on the nervous system, this difficulty is compounded by the difficulty in obtaining good translations of the majority of the previous work.

REFERENCES

Ely, T. S., D. L. Goldman, 1956, Heat Exchange Characteristics of Animals Exposed to 10-cm Microwaves, IRE Transactions - Medical Electronics, February, pp. 38-43.

Gvozdikova, Z. M., V. M. Anan 'yev, I. N. Zenina, V. I. Zak, 1964, Sensitivity of the Rabbit's Central Nervous System to Continuous Superhigh, Frequency Electromagnetic Fields, Byulleten Eksperimental 'noy Biologii i Meditsiny, 8:63-68.

Ingalls, C. E., Personal Communication.

Kamenskiy, Yu I., 1964, The Effect of Microwaves on the Functional State of a Nerve, Biofizika, Vol. 9, No. 6, pp. 695-700.

Michaelson, S. M., R. A. E. Thomson, J. Q. William, 1967, Effects of Electromagnetic Radiations on Physiologic Responses, Aerospace Medicine, March, pp. 293-298.

Schwan, H. P., Kam Li, 1956, The Mechanism of Absorption of Ultrahigh Frequency Electromagnetic Energy in Tissues, as Related to the Problem of Tolerance Dosage, IRE Transactions-Medical Electronics, February, pp. 45-49.

Turner, J. J., 1962, The Effects of Radar on the Human Body (Results of Russian Studies on the Subject), Armed Services Technical Information Agency No. AD278 172, pp. 22-24.

APPENDIX

A. From a Translation by Turner:

   Numerous Russian works as well as those of foreign authors including Barron, Love, Daily, Hines, and Randall of the U. S. A. give evidence of the fact that functional changes of the nervous system are observed in persons who have been subjected to microwave irradiation. These works do not, however, consider the intensity of the microwave radiation to which the irradiated persons were exposed.

   Polyclinical and clinical observations of the health of 525 persons (352 men and 173 women) were conducted at the Institute of Labor Hygiene and Occupational Diseases of the Academy of Medical Sciences, U. S. S. R. over a period of several years. These persons were regulators, assemblers, and engineering-technical workers servicing various microwave generators. All had been exposed to microwave radiation of various intensities and durations.

The exposed personnel were divided into three groups according to the intensity and duration of the microwave radiation to which they had been subjected.

Group I - (184 persons) included persons periodically exposed to intensive microwave irradiation (up to several milliwatts per square centimeter).

Group II - (263 persons) included persons periodically exposed to low-intensity microwave irradiation (up to 1 milliwatt per square centimeter).

Group III - (78 persons) included persons periodically exposed to microwave irradiation of still lower intensity (up to tenths of a milliwatt per square centimeter).

The length of employment of those examined varied from several months to five to ten years and more however, most of them had been employed for four years (65.9 per cent). One hundred teachers of higher educational institutions, of approximately the same age group (25 to 35), served as the control group.

Among the persons exposed to the microwave energy, asthenic symptoms accompanied by vascular-vegetative changes and certain endocrine disorders were frequently observed. Manifestation of the symptoms depended directly on the intensity and duration of the irradiation.

Most of those examined complained of increased fatigability, periodic or constant headache, extreme irritability, and sleepiness during work (Table I).

Table I

Group	Number examined	Changes in the Central Nervous System (per cent)									
		1	2	3	4	5	6	7	8	9	10
I	184	12	20	8	2	16	73	19	27	6	15
II	263	39	35	27	12	10	24	4	28	28	37
III	78	36	31	15	19	14	38	11	32	26	61
Conrol	100	8	10	8	2	-	3	-	14	4	14

Key to Table I
1. Headache   6. Bradycardia
2. Increased fatigability   7. Slowing of the OKSR
3. Increased irritability   8. Arterial hypotension
4. Sleepiness   9. Hyperhidrosis of the wrists
5. Inhibited dermographia   10. Enlargement of the thyroid gland

B. From Gvozdikova

Distribution of Number of Cases of Various Changes in EEG According to PPM of Field

Wave length, cm	PPM, mw/cm²	Number of cases with changes in EEG SVCh Field				Number of cases without change in EEG	Relationship of number of cases with changes in EEG to total number of irradiations, %
		drop in frequency with increase in amplitude of basic rhythm	increase in frequency of basic rhythm with amplitude changes on both sides	drop in amplitude without change in frequency of basic rhythm	in total		
12.5	0.02	4	2	2	8	4	67
	0.08	2	1	1	4	7	36
	0.4	1	4	-	5	6	45
	2	5	3	-	8	2	80
	10	2	3	3	8	2	80
	50	5	1	1	7	2	78
52	0.02	5	5	2	12	2	86
	0.08	3	1	2	6	5	55
	0.4	2	5	2	9	5	64
	2	7	1	-	8	2	80
	30	3	5	1	9	1	90
	50	7	5	-	12	2	85
100	0.02	7	3	-	10	1	90
	0.08	1	2	1	4	6	40
	0.4	6	3	-	9	1	90
	2	4	3	2	9	3	75
	10	7	4	2	18 (sic)	1	93

EFFECT OF CORTICALLY APPLIED DC CURRENTS ON STRYCHNINE SPIKE ACTIVITY - A PRELIMINARY REPORT

H. Fukuda , H.J. Marsoner and F.M. Wageneder, M.D.

Chirurgische Universitaetsklinik, Graz, Austria

INTRODUCTION

It is well known, that a steady potential is to be observed between the cortical surface and inactive parts of the brain, like the ventricles (O'Leary, J.L. & Goldring 1959). In a paroxysmatic convulsive state, however, the negativity of the cerebral cortex changes to positivity. However, the negative potential of the cortex increases with a spreading depression (Harreveld & Cchade 1962). It is seen that this steady potential can be influenced in many ways, e.g. a repeated low frequency stimulation (6 to 10 cps) of the median or the lateral nucleus of the thalamus, creating a recruiting response or an augmenting response on the cerebral cortex also shifts the cortical d.c. potential to more negative values. (Caspers 1959). The same effect can be obtained by repeated stimulation of the reticular activating system (Caspers 1959).

Considering the facts briefly reviewed above we can expect a large influence of diffusely applied d.c. currents on the brain activity.

Looking for a simple indicator of the neuronal activity we decided to use the Strychnine-spike for this study. Strychnine is a thoroughly examined drug and, as widely assumed, it cancels the postsynaptic inhibition.

Method

The experiments were performed on 20 adult rabbits regardless of sex. Their weight was between 2.3 and 2.5 kg. To prevent brain

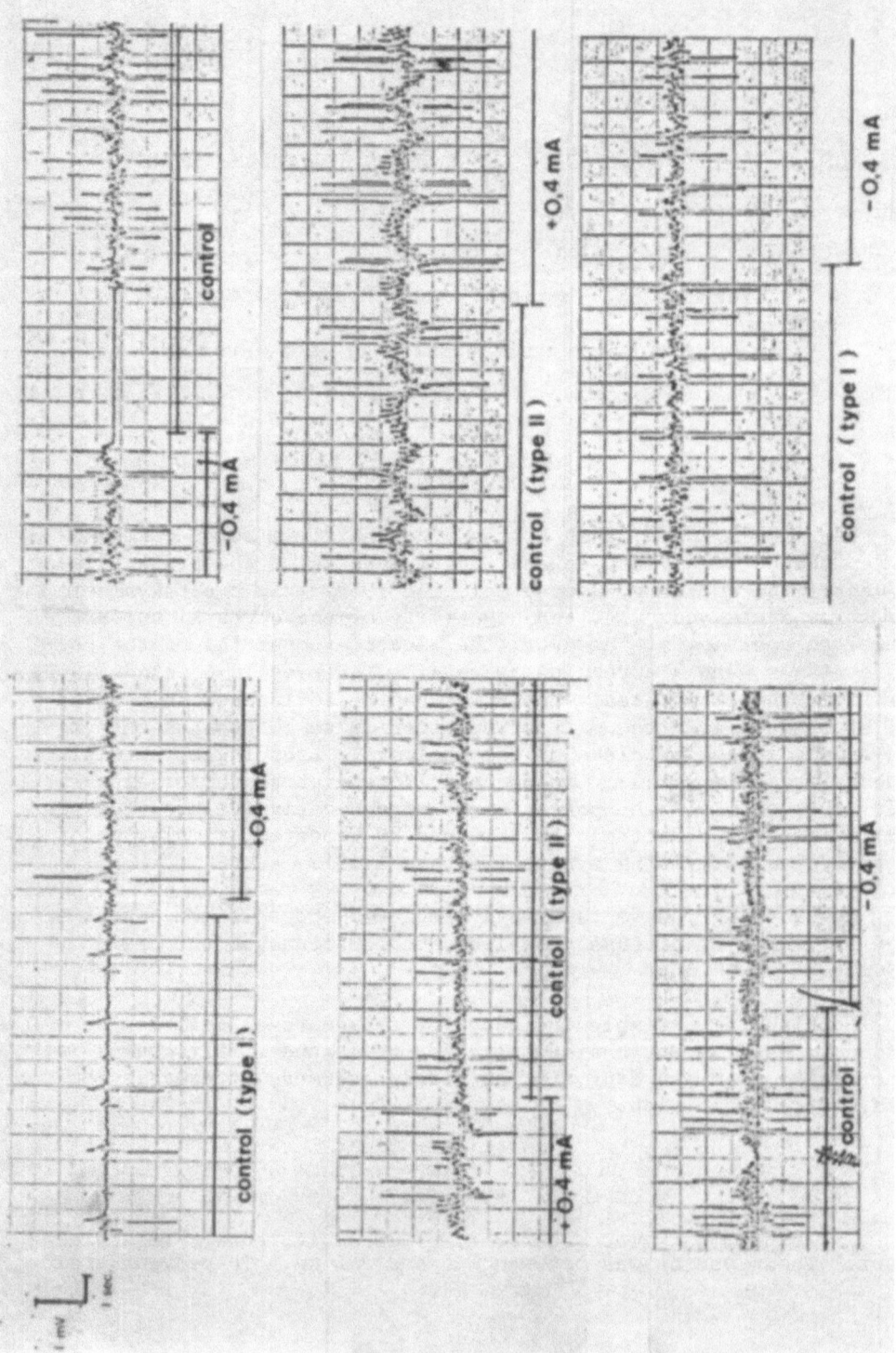

Figure 1

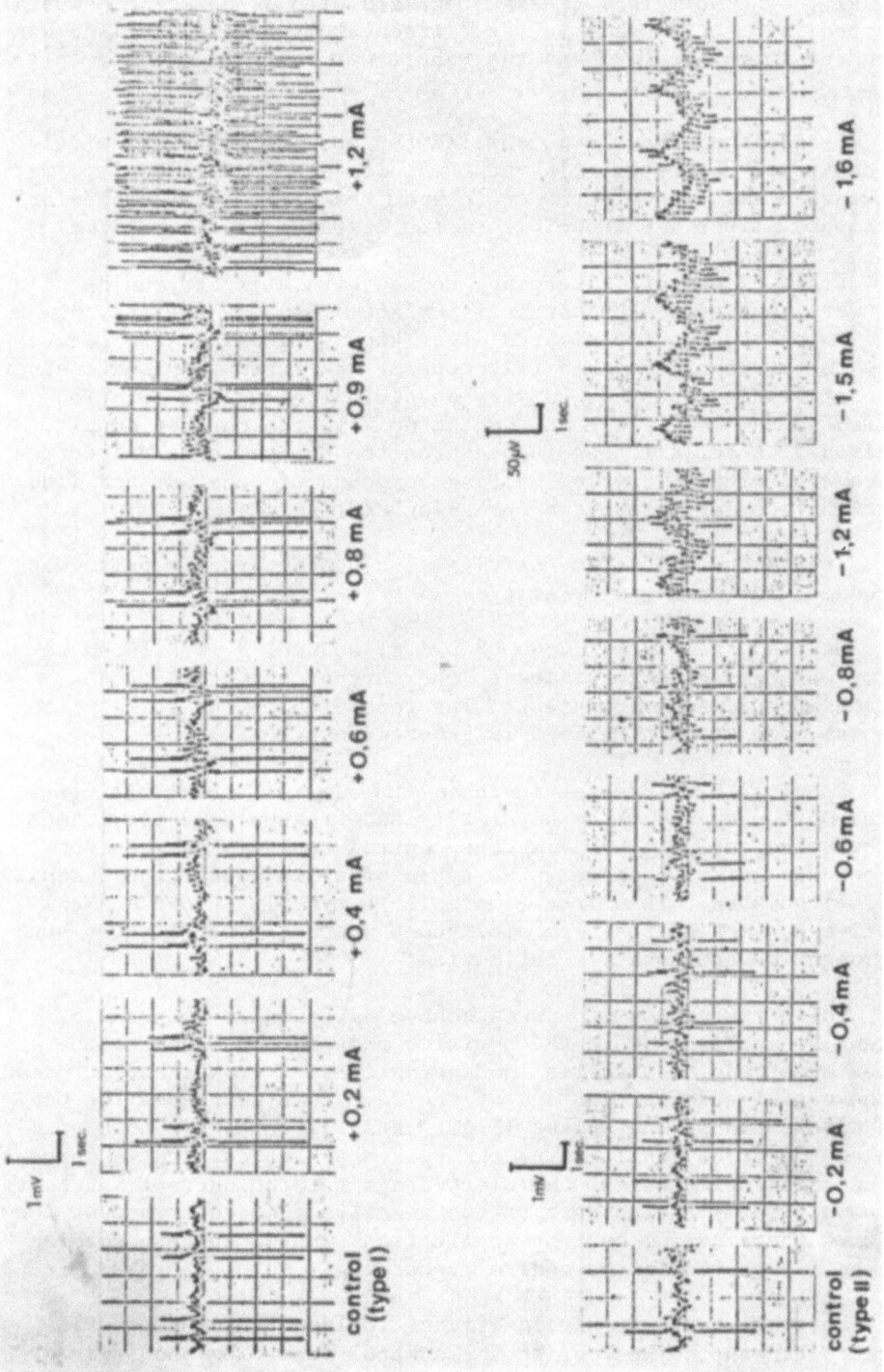

Figure 2

edema 10 to 15 ml of a 60% glucose solution were injected intra-
venously. The animals were anaesthetized with 30-35 mg/kg Nembutal
injected intraperitoneally. After tracheotomy Succinylcholine was
given for immobilisation and the rabbit was connected to a respir-
ator.

Standard procedure was applied for opening the skull widely.
The center of the opening lay at AP 5; its length was about 12 mm,
the width 8 mm. The electrocorticogram was recorded by a bipolar
balltipped silver electrode with a tip distance of about 10 mm.

Small pieces of filterpaper soaked with a 1% Strychnine solution were applied to the cortical surface. Usually the strychnine
spikes appeared about 3 minutes after the application; two minutes
later the strychnine soaked filter paper was taken away and another
piece of filter paper soaked with physiological saline solution
applied to the same place and contacted with the current supply
electrode. After all these procedures the exposed cerebral cortex
was covered with a very low melting compound of bone wax and liquid
paraffin in order to avoid a spreading depression.

The second electrode, serving as a d.c. current supply, was
introduced into the nasal cavity.

We used an OLTRONIX C 40-08 D d.c. source; a 10 kiloohm resistor was connected in series to the current path to keep the
current approximately constant. The recording of the spike potential was done using a Hellige 19 EEG-recorder.

Results. Our interest in these experiments was mainly concentrated on the change of both polarity and frequency of occurrence
of the spike activity. During the control recordings, when the
current was not yet applied, two types of spike patterns can easily
be distinguished. The first - we call it type I - shows a small
negative but a high positive amplitude, whereas type II shows both
high negative and positive deflection.

These two groups of spikes behave differently when a d.c.
current is applied. When the positive pole is connected to the
cortex electrode, a stepwise increasing current evokes an increase
of the negative deflection and at the same time decreases the positive elongation of the spikes (Fig. 1 and 2). In Fig. 3 both positive and negative amplitude of the spikes are calculated in percent
of the total spike height and plotted against the current intensity.
It shows clearly an increase of the negative and a decrease of the
positive phase during current-application, when a spike according
to type I appeared in the control recording.

However, there is no significant influence of the positive
current when the spikes during the control recording show already

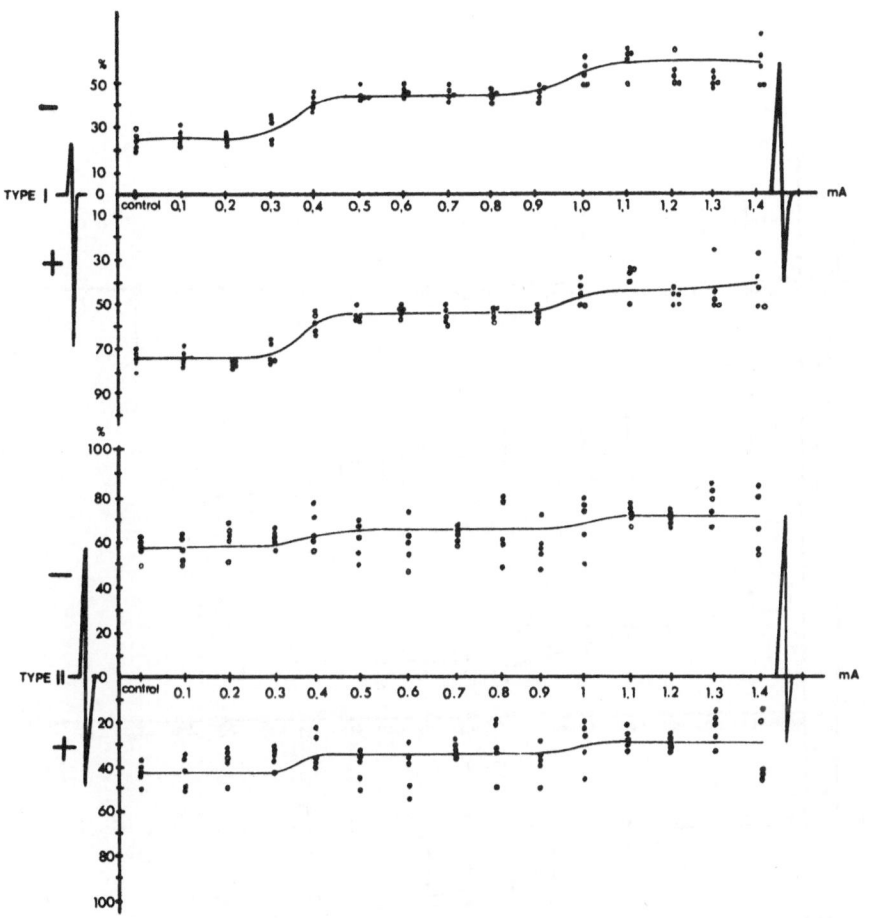

Figure 3

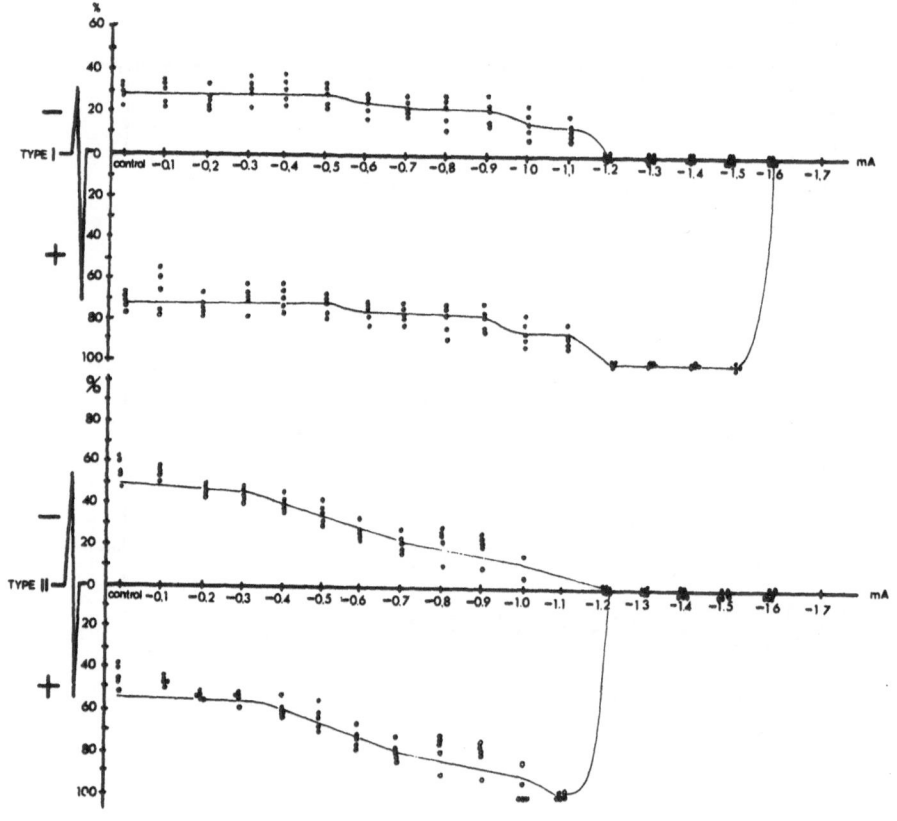

Figure 4

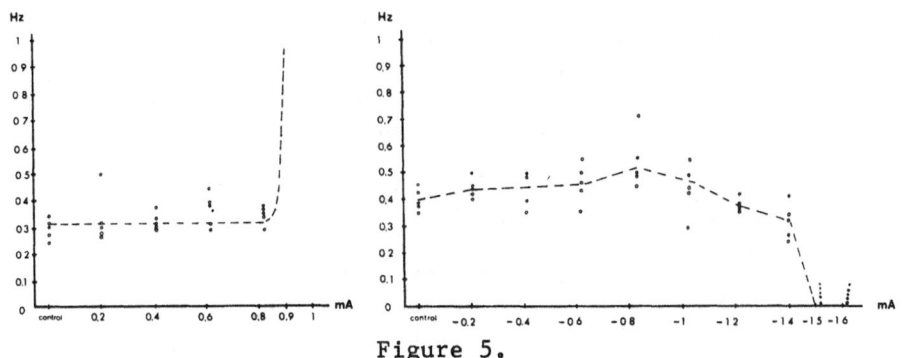

Figure 5.

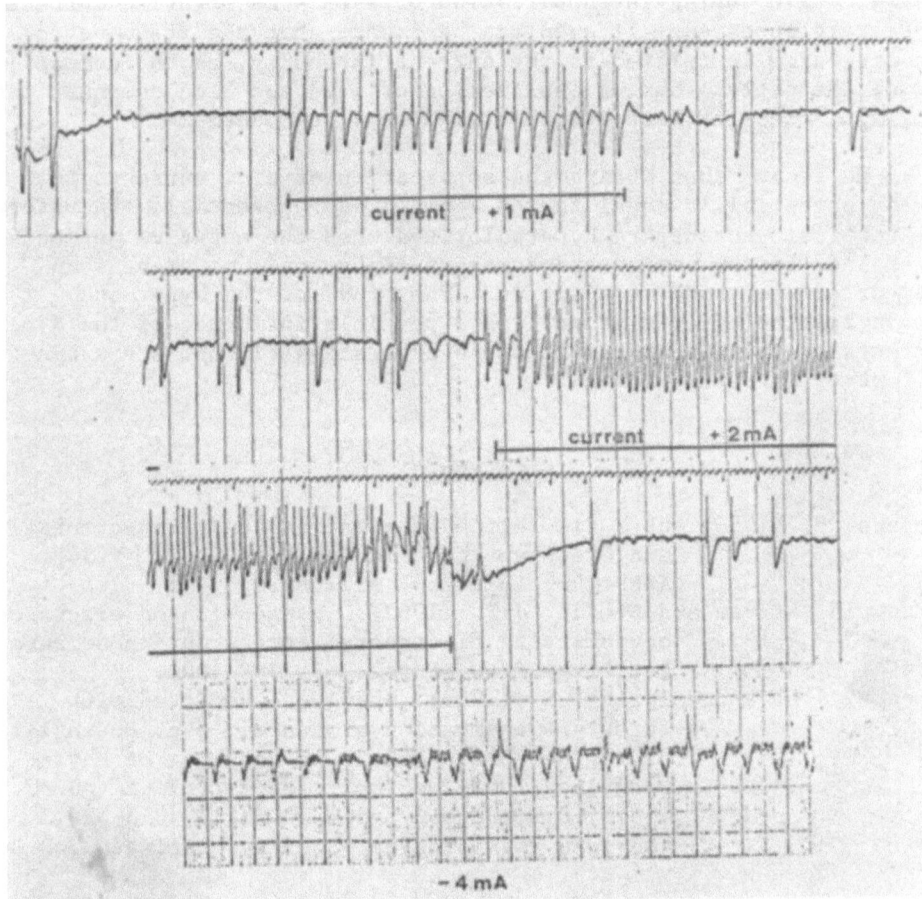

Figure 6.

almost equal elongations in both directions (type II).

The result is quite different, when the negative terminal of
the current source is connected to the cortex-electrode. Then the
negative phase of both types of control-spikes decrease; gradually,
but at the same time, the total amplitude decreases (Fig. 4). At
a current level of 1.2 to 1.6 mA the spike disappears.

The lower part of Fig. 5 shows examples of the recording
application of different current intensity.

The same figure demonstrates that there is another important
difference whether the positive or the negative terminal of the
power source is connected to the cortex.

A positive current of 1.2 mA always produced seizure discharge, while during the application of even 4 mA negative current
no seizure appeared. It even can be found, that the frequency of
the discharge is related to the current intensity, as the comparison of the pattern during application of 1 mA and 2 mA demonstrates
(Fig.5). These effects are reversible and reproducible.

We assume that during the application of d.c. currents the
steady potential is equal to the applied d.c. potential. Therefore,
the cortical neurons are hyperpolarized when the negative potential
is applied to the cortex and depolarized by ionic movement when
the cortex electrode is positive. The study of the hyper-and
depolarization effects as well as a possible influence of the d.c.
currents on the thalamo-cortical reverberating circuit are subject
to further investigation.

Literature

Caspers, H. (1959): Uber die Beziehung zwischen Dendritenpotential
                    und Gleichspannung an der Hirnrinde. Pflug.
                    Arch. Physiol. 269; 157-181.
Harreveld, A. van and Schade, J.P. (1962): Changes in the electrical
                    conductivity of cerebral cortex during seizure
                    activity. Exp. Neurol., 5, 383-400.
O'Leary, J.L. and Goldring, S. (1959): Changes associated with
                    forebrain excitation processes: d.c. potentials
                    of the cerebral cortex, in Handbook of Physi-
                    ology Section 1 Neurophysiology Vol. i (ed. J.
                    Field, H.W. Magoun and V.E. Hall), 315-328,
                    American Physiological Society Washington, D.C.

PARKINSON-LIKE TREMOR PRODUCTION BY TRANSCRANIAL STIMULATION

R.S. Pozos and J.R. Holbrook

Departments of Physiology and Anatomy

University of Tennessee Medical Units

The mode of anesthesia and sleep production by electrical current remains an enigma. Numerous reports have appeared on satisfactory anesthetic states produced by wide range of frequencies, various levels of current and different types of electrodes(1,2). The attraction of electroanesthesia(EA) is the immediate recovery from the anesthetic state once the electrical current is off. Electrosleep(ES) offers a potential alternative to electroconvulsive therapy due to its low current levels (3). However, widespread use of these procedures must be guarded until the mechanism of action is more clearly elucidated.

A possible model of EA and ES was proposed based on the production of a prolonged Parkinson-like state in reserpinized dogs (4). By combining low levels of reserpine (.25 mg/kg) and EA (30 ma,4 v, 700 hz) a Parkinson-like state was produced. The dogs demonstrated ptosis,flexor dominance, rigidity and tremor. This state persisting for five to seven days could be reversed by atropine, L-dopa or EA. The reversal with electrical current persisted for three to four minutes after electrical stimulation. Cholinergic drugs (neostigmine, physostigmine) would exacerbate the syndrome. Electroshock and ES also produced the same state in reserpinized dogs. Animals subjected to EA or given reserpine (.25mg/kg) did not develop this syndrome. Complete details on this procedure have been published (3,4). Since electrical current in conjunction with reserpine caused a dopamine depleted

animal(Parkinson-like) it was proposed that the mode of
action of electrical current in the normal dog might be
due to an inhibitor(catecholamine) efflux in the central
nervous system. This theory is somewhat substantiated
since epinephrine injected intracerebrally produces an
anesthetic state(5). Also DL-dopa and 5-Ht raises the
nocioreceptive threshold in rats when injected intra-
muscularly(6).

Though a Parkinson-like dog had been produced the
similarities between the animal model and the Parkinson
patient could not be determined. Recently, Parkinson
patients undergoing L-dopa therapy have become available
for such comparison studies(7).

METHODS

An accelerometer was taped to the index finger of
Parkinson patients whose forearm was supported by a cradle.
Dogs were placed in a sling which allowed their legs to
swing freely and an accelerometer was taped to the distal
part of the paw. Control observations were made after
which dogs were given EA. Finally EA was given to
reserpinized dogs.

Histological investigation of the nucleii of the
extrapyramidal system was begun in the Parkinson-like
dog since these nucleii are affected in Parkinson-patients.
Dogs were anesthetized and exsanguinated by viviperfusion
with saline followed by formalin. Brains from control dogs,
reserpinized dogs and Parkinson-like dogs (reserpine and
EA) were removed and placed in 10% formalin. After appropriate fixation time, the brains were embedded in paraffin.
The mesencephalon was sectioned at six microns and stained
with thionin and hematoxylin and eosin. Brains from experimental were removed four hours after the injection
of reserpine or reserpine and EA. This amount of time
is required for the reserpinized dog to show maximal
effects of the drug. Dogs given reserpine and EA demonstrated Parkinson-like signs one-half an hour after
stimulation.

RESULTS

It was difficult to contain an unanesthetized dog
in the sling for any period of time. Further, once the
dog was conditioned to stay in the sling, shivering would

ensue. In three dogs conditioned to stay in the sling and not shiver a resting limb tremor of 10 hz was recorded. This value is contrasted from the shiver frequency which is approximately 15 hz. Upon EA stimulation the limb tremor recorded was 5-6 hz. After stimulation the normal frequency returned. In the reserpinized dog given electrical current the limb tremor was again 5-6 hz. However this low frequency would remain after electrical stimulation until the other signs of Parkinsonism abated (5-7 days). Parkinson patients have a 5-6 hz tremor which was similar to that seen in the dog.

Histological study of the substantia nigra is now in progress. In all experimental animals, the neurons of the substantia nigra are enlarged and vacuolated. Nissl substance in many neurons appears clumped. Some neurons have the appearance of excessive Nissl clumping at the neuronal poles. Further the structure of the majority of the cell population shows a change from typical multi-polar configuration to one of oval and elongated shapes. These changes are most marked in the Parkinson-like dog. Such alterations are present to a far lesser extent in the reserpinized dog who received no electrical current. Neurons of the substantia nigra in control animals showed none of the above changes.

Additional Comparisons between Parkinson-like Dog and Parkinson Patient

Creeping is one of the exercises recommended for Parkinson patients. It is a common observance for the patient to creep for a short period of time and then to slowly fall to one side due to weakness in the arms (8). The Parkinson-like dog is also unable to walk for any period of time and demonstrates forelimb weakness.

A considerable number of Parkinson patients suffer from or have a past history of gastrointestinal disorders (9). Chewing difficulties and drooling are also common (10). The Parkinson-like dog demonstrates chewing dif-ficulties and bloody stools.

DISCUSSION

The production of the Parkinson-like tremor in dogs with transcranial currents implicates the mesencephalon as being one of the brain areas involved in EA(11). Stimulation of the substantia nigra (mesencephalon) would

cause modification of the levels of neurotransmittors
of the striatum by way of the nigro-neostriatal tract
(12,13). The caudate and putamen have the highest levels
of dopamine(14) and acetylcholine in the central nervous
system(15). Depletion of catecholamines(electrical current
and reserpine) would cause a dominance of cholinergic
substances producing some of the characteristic signs of
Parkinsonism.Stimulation of these areas after the animal
is Parkinsonized would explain the temporary reversal
of the Parkinson-like state due to a temporary increase
in catecholamines and/or other inhibitors. This hypothesis
is consistent with current reports of catecholamine involve-
ment in anesthesia(5), sleep(16) and electroshock(17,18).
Whether other inhibitory substances(serotonin,gaba) are
involved in the action of electroanesthesia cannot be
determined from these present studies. The possibility
of ion efflux determining catecholamine secretion in EA
and electroconvulsive shock has been considered(3).
Implication of the thalamus as being the main target site
of electroanesthesia(19) is consistent with what has been
proposed,since stimulation of the thalamus causes dopamine
secretion in the caudate. By means of the mammilothalamic
and the mammilotegmental tracts the hypothalamus is intimate-
ly related to the thalamus and midbrain. Thus structures
involved in anesthesia(thalamus) and sleep(caudate[20],
thalamus[21] and hypothalamus[22]) are anatomically
connected.

    Interestingly enough there is anothercommon anatomical
denominator of these structures;namely the ventricular
system. Driscoll has shown that the presence of a theoretical ventricle would modify the current density(23).
Marsoner has shown that the injection of a dielectric substance into the ventricles increases the current requirements for EA by 24-47%(24). Other investigators have
also proposed the ventricles as probable sites in the
production of sleep and anesthesia(25).

    Transcranial stimulation has attracted our attention
in terms of catecholamine(inhibitor) involvment. However
the interaction of the cerebrovascular system and electric
current should not be precluded. Some of the changes
observed in the Parkinson-like dog's substantia nigra are
similar to those reported for hypoxia(26). Electroshock
is also reported to cause similar neuronal changes(27).
However since these changes are reversible they should
not be considered pathological(27). Other investigators
have shown that blood flow in the brain increases during
REM and Slow Wave Sleep(28). At present whether the primary
action of electrical current is on the neuronal membrane

and secondarily on the cerebrovascular system cannot
be determined.

Acknowledgment: This work was supported in part by
grants FR-5423 and HE-05612, National Institutes of
Health, U.S. Public Health Service.

BIBLIOGRAPHY

1. R. H. Smith, J. Tatsuno and R. Zouhar,Anesth.
    and Analg. 46(1),109 (1967)
2. R. A. Herin, Am.J. Vet. Res. 29(3), 601 1967
3. R.S. Pozos, L.E. Strack, R.K. White, And A.W.
    Richardson, Neuroelectric Conference 5, 17(1969)
4   R.S. Pozos, A.W. Richardson, And H. M. Kaplan,
    Anesth. and Analg. 48(3) 342 (1969)
5. A. B. Rothballer, Pharmacol Rev. 11, 494 (1959)
6. S. Radouco-Thomas, P. Singh, F. Garcin, and C.
    Radouco-Thomas, Arch. Biol. Med. Exper. 4, 42 (1969)
7. Permission of Dr. R. A. Utterback, Professor and
    Chairman Department of Neurology, University of
    Tennessee Medical Units to participate in L-dopa
    Research Study on Parkinson Disease is gratefully
    acknowledged.
8. E. Randolph, Director of Physical Therapy,Baptist
    Memorial Hospital, Memphis Tennessee.
9. R. R. Strang, Acta Neurol. Scand. 42(1),124 1968
10. M.J. Eadie, and J.H. Tyrer, Aust. Ann. Med. 14,13(1965)
11. E.A. spiegel, H.T. Wycis, E.G. Szekely, B. Rusy,
    and H.W. Baird, Arch. Neurol. 2, 56 (1960)
12. A. Dahlstrom and K.Fuxe, Acta. Physiol Scand.62,232(1964)
13. N.E. Anden, A. Carlsson, A. Dahlstrom, K. Fuxe,
    and K. Larsson,Life Sci. 3, 523 (1964)
14. O. Hornykiewicz, Pharmacol. Rev. 18(2) 791 (1966)
15. F. C. MacIntosh, J. Physiol. 99, 436 (1941)
16. M. JOuvet, Science 163, 32 (1969)
17. P. Schubert, and W. Landisisch, Psychopharmacol.
    (Berl) 15, 289 (1969)
18. R. Kaebling, E.G. Koski and C. O. Hartwig. J. Psychiat.
    Res.6, 153 (1969)
19. D.V. Reynolds, Science 164, 444 (1969)
20. R.G. Heath and R. Hodes, Tr. Am. Neurol. 77, 204 (1952)
21. W. R. Hess, Helv. Physiol. Acta. 2, 305 (1944)
22. G. Maresco, O. Sager, and A. Dreindler. Ztschr. ges.
    Neurol. Psychiat. 119, 277 (1929)
23. D. Driscoll and S. Rush, Neuroelectric Conf.5, (1969)
24. J. J. Marsoner, J. Tatsuno, and F. M. Wageneder.
    Neuroelectric Conf. 5, 58 (1969)

25.   G. M. H. Waites, J. Physiol. 139, 417 (1957)
26.   D.E. Fletcher,J. Neuropath and Expt. Neur.6,299 (1947)
27.   K.T. Neurbuerger, R.W. Whitehead, E.K. Rutledge and
      F.G. Ebaugh. Am. J. of Med. Sci. 204, 381 (1942)
28.   M. Reivich, G. Isaaca, E. Evarts and S. Kety.
      J. Neurochem. 15, 301 (1969)

ELECTROSTIMULATION OF HEARING [1,2]

Michael S. Hoshiko, Ph.D.

Department of Speech Pathology and Audiology

Southern Illinois University, Carbondale, Ill. 62901

Volta in 1800 first reported evoking hearing sensation by direct electrical stimulation. Since that time many experimenters have produced a hearing sensation by placing one electrode in the ear canal which had been filled with a conducting fluid and the other electrode on the arm. This phenomenon of hearing was named "electrophonic hearing" by Stevens (1937). Stevens removed the wires from a speaker of a radio set and connected electrodes to the wires. The result was highly distorted music and speech which was not intelligible. By adding DC polarizing voltage both music and speech became "dramatically clear" compared to the previous report of "a mere sequence of sounds." Davis and Silverman in 1962 stated that electrophonic research had practically ceased because it offered so little promise.

Recently, a second method for electrostimulation of hearing using amplitude modulated radio frequency waves has been reported which seems more promising than direct electrical stimulation and will be called "radiophonic hearing."

The phenomenon of hearing electromagnetic radiations may have been reported as early as 1783 when Sir Charles Blagden collected all available reports of persons hearing hissing sounds while watching meteorite fireballs plunge to earth which, according to Romig and Lamar (1963), generate electromagnetic waves. Stevens and Davis in 1938, briefly mentioned that Stevens and Hunt had used a 400 Hz modulating tone with a 100K Hz radio frequency wave and were able to demonstrate that subjects could "hear" the modulating tone. However, this line of research was not pursued.

An intensive search of the literature has indicated that most
of the investigators were concerned with the effect of electromagnetic radiation on biological tissue damage, and that they were
motivated after the development of high powered radar transmitters.
However, some investigators reported various athermal effects.

In Italy as early as 1930, Nrunori and Torrisis reported that
humans can detect electromagnetic energy. Manfredi and his
collaborators have been investigating radiophonic hearing as early
as 1951. In 1963, he reported hearing enhancement for the deaf
through radiophonic hearing stimulation. Radiophonic research has
also been reported by Russian investigators. In 1949, Sheyvekhman
reported that radiophonic stimulation increased hearing thresholds.
Pressman (1963) reported his subjects both normals and deaf
"heard" unmodulated electromagnetic radiation as whistling, humming
and clicking sounds.

Frey (1962) reported that his subjects, normals as well as
hard of hearing and clinically deaf, responded in at least a part
of the radio frequency spectrum. Constant (1967) reported that his
subjects "learned" to hear electromagnetic waves.

Moeser (1964) reported that a high school student named
Flanagan had "invented" a new hearing device. At about the same
time, Puharich and Lawrence (1964) reported that they were able
to stimulate hearing in the deaf with amplitude modulated radio
frequency waves. After the appearance of these reports, Sommer
and Von Gierke; Harvey and Hamilton, under the supervision of
Sommer and Von Gierke concluded that the hearing was not unlike
bone conduction. This conclusion was questioned by Frey (1968)
who based his evaluations upon mathematical calculations of the
pressures produced by electromagnetic radiation pressure specifically
developed for biological material which Sommer and Von Gierke,
Harvey and Hamilton had failed to do. Privately, Sommer (1968)
stated that he felt that their conclusions may need revision.
Puharich and Lawrence have conducted a rather extensive investigation
on radiophonic hearing. However, their results have not been
readily available.

In summary then, it was felt that electrostimulation of hearing by amplitude modulated radio frequency waves should be
investigated. Therefore, arrangements were made to test the hearing
devices developed by Flanagan and also by Puharich and Lawrence.

On the instrument developed by Flanagan subjects heard tones
from 500 Hz to about 20,000 Hz at equivalent to about 20dB in
loudness. Some subjects could hear down to about 250 Hz and none
below that frequency. Music and speech could be heard although
connected discourse could be followed only with great effort. The

instrument worked intermittently and it had to be returned for repairs several times. However, it did induce hearing even though poorly.

The rest of the paper is based upon the results of the second instrument.

A total of 29 college students with normal hearing were tested with the hearing device developed by Puharich and Lawrence. The hearing device was modulated by the output of a Hewlett-Packard 207A oscillator, a Fisher FM tuner and from the CID W22 record. Two electrodes one inch in diameter insulated with mylar were applied on the skin in the area just in front of the tragus. They were held in place by a head band not unlike that used for ear phones.

Testing was done with the CID recorded W22 phonetically balanced words. 20 Ss were tested. The total number of errors was 91 from a total of 1000 words. The average number of words missed was 4.5 for the list of 50 words. The range was zero missed to a high of 8 words. The average discrimination was therefore 90%. Since over half of the subjects missed the same word and one word was missed by 19 of the subjects it was felt that the scores were depressed because of the condition of the records. Nine additional subjects were tested using new records. In addition, the testing was conducted in the I.A.C. anechoic chamber. Under the more stringent testing conditions the discrimination scores did improve. The subjects were first tested with radiophonic stimulus then with acoustic stimulation using the same list. It was felt that under this testing order all of the advantages would go to the acoustic stimulation. For the radiophonic presentation the results were as follows: from the total of 450 words, 28 words were missed by the Ss. The average number of words missed for the list was 3.1. Under the acoustic conditions the total number of words missed was 22 and the average number of words missed per list was 2.3. The discrimination scores were 94 and 96% respectively. It was interesting to note that four same words were missed more frequently in both conditions and three of the words were missed more frequently in the acoustic condition. The t test between the two conditions (.24) was not significant. Therefore, it was concluded that the discrimination scores under the two conditions were not different.

With the instrument connected to the FM tuner, music and speech were heard with the same quality as from an acoustic system. There was no difficulty in understanding news broadcasts. Subjects heard pure tones from 30 Hz up to about 20,000 Hz equally well at 30dB loudness ASA determined by loudness balancing technique.

Other research with radiophonic stimulation is in progress.
It appears that radiophonic stimulation of hearing may open up
other areas for research in audition since it eliminates all of
the disagreeable side effects induced by electrophonic stimulation
and the quality is the same as with acoustic stimulation.

Selected References

Frey, Allan H., "Human Auditory System Response to Modulated
      Electro-magnetic Energy," J. of Applied Physiology,
      pp. 689-692, Vol. 17, 1962.

Manfredi, Angelo, Terapio radioacoustica della sordita. La
      Ricerca Scientifica, Anno 33, Vol. 33, No. 3, pp. 217-228,
      1963.

Moeser, William, "Whiz Kid, Hands Down," Life Magazine,
      pp. 69-70, September 14, 1962.

Puharich, H. K. and Lawrence, J. L. "Modulated alternating cur-
      rent energy used to stimulate audition in totally deaf
      humans." Paper presented at the 25th Annual Meeting of the
      Aerospace Medical Association, Miami, Fla., May 11-14, 1964.

Romig, Mary F., Lamar, D. L. "Anomalous Sounds and Electromagnetic
      Effects Associated with Fireball Entry," Memorandum
      RM-3724-ARPA, Advanced Research Projects Agency, July 1963.

Sommer, H. C., Von Gierke, H. E., "Hearing Sensations in Electric
      Fields, Aerospace Medicine, Vol.35, No. 9, pp. 835-839,
      September 1964.

Stevens, S. S., "On Hearing by Electrical Stimulation." Journal
      of the Acoustic Society of America, pp. 191-195, Jan., 1937.

1. The investigation was initiated when the author held a Special
    Post-doctoral Fellowship from the National Institute of Health
    (1 F10 NB 1598-01 C D R) while on leave to the Neurocommunica-
    tion Laboratory, John Hopkins University, School of Medicine.

2. The author wishes to thank Mr. Warren Austin of Borders
    Electronics for making available the second instrument for
    test purposes.

GRADABLE PAIN PRODUCTION

Robert H. Smith, M.D. and J. Herbert Andrew, M.D.

Department of Anesthesiology, University of California

Medical Center, San Francisco, California

We are engaged in a continuing inquiry into the possibility of producing clinical anesthesia solely by the application of electrical current.

One of the needs of the inquiry has been a harmless means of producing pain against which we might "measure" anesthetic potency. We decided that pain would, to be satisfactory, have to have the following characteristics:

1)   A severity comparable to a surgical incision, or, even better, comparable to a muscle splitting incision.

2)   Controllable as to severity so pain intensity could be decreased or increased as desired.

3)   Ends instantly when desired.

4)   No tissue injury produced.

5)   Repeatable with good degree of uniformity of results in repeated tests.

6)   No "startle" component such as is seen with electric shock.

Our present and most satisfactory means of gradable pain production involves a specific instrument and a pattern of application.

The instrument is a pulse generator with the following characteristics:

1)   The pulse form is the bidirectional "square" pulse with a
very sharp leading edge.   The duty cycle is variable over a range
from 1% to 48%.

2)   The negative pulse can be oriented anywhere between the end
of one positive pulse and the start of the next one.

3)   The frequency can be varied from 2 per second to 1,000 per
second.

4)   Peak milliamperage range is 0 - 100 mA.

5)   The pulse duration (same for positive and negative at a given
setting) can be varied between 5 micro seconds and 5 milliseconds.

6)   Incorporated in the generator is a device which will deliver
a 10 mA linear increase in current intensity in 6-½ seconds then
drop the intensity back to the starting point in 0.3 seconds.   The
10 mA increase can be added to any current level below 90 mA.   We
call this "ramping in" 10 mA.

        The instrument output was designed to meet the characteristics
of the pain pattern we sought.   The bidirectional current avoids
tissue injury by iontophoresis.   Large saline-soaked sponge electrodes prevent burning by current density.   The pain severity is a
function of milliamperage which is readily controllable.   "Ramping
in" of current increase prevents the "startle" response of suddenly
applied current.   The pain produced is a combination of muscle
spasm, skin response, and a nerve stimulation best described as
tingling.   Bluntly speaking, the instrument produces a form of pain
which is simply intolerable.   When the subject reaches his "intolerable point" he feels as though the tissue is being ripped from
the bones of his forearm.   We have not tried to compare the pain
with that of an incision.   Instead, we simply apply increasingly
severe pain until the subject gasps and demands relief.

        The pattern of application is simple.   We apply two large
sponge electrodes to the anterior and posterior aspects of the
subject's forearm at about the midpoint.   The sponges are taped in
place.   The most effective pattern is a frequency of 25 pulse pairs
per second with a pulse duration of one millisecond, the negative
pulse midway between the two positive pulses, and the duty cycle
about 5%.   Starting at zero mA, the current is built up slowly over
two to four minutes to the subject's "intolerable point".

        At that point the current intensity is rapidly reduced 10 mA.
The sudden decrease is a tremendous relief despite the fact that
the muscles of the forearm are still in spasm.

        Testing the effeciency of any anesthesia-producing procedure

can begin at this point. When a test is indicated, 10 mA are
"ramped in" over 6-½ seconds. This raises the current to the
subject's "intolerable point", but drops back 10 mA immediately.
If the subject's response to the "intolerable level" is modified,
the milliamperage can be increased to ascertain his new "intolerable
level." The increase in mA is some sort of a measure of the analgetic effect of whatever is being tested.

A few specific points:

Subjects can not tolerate the "intolerable point" for more
than a few seconds. The pattern of application being described
permits the "intolerable point" to be reached as desired for a very
short time to test analgesia. There is no "startle" effect as is
seen with simple electric shock. The amount of current to produce
the "intolerable point" seems to correlate directly with the diameter of the arm. The range in eight subjects in this laboratory
was from 28 mA to 48 mA. Subjects will tolerate current intensities
10 mA below their "intolerable point" for up to 30 minutes despite
the fact that their forearm muscles are in severe spasm.

The "intolerable point" is a constant for a given subject with
a given current pattern and electrode location. In successive tests
the point will seldon vary more than one or two milliamperes.

Electrodes should have one square inch of contact to avoid any
risk of current density burns.

Acknowledgement

This research was supported by the John A. Hartford Foundation.
The authors gratefully acknowledge the use of facilities at the
San Francisco General Hospital (San Francisco, California) and the
technical assistance of Peter A. Lindquist, Robert H. Dempsey, and
Mrs. Shelley Frisch.

THE USE OF APPLIED DC FIELDS IN THE ANALYSIS OF INTERICTAL

EPILEPTIFORM DISCHARGES

Curtis A. Gleason

Department of Neurology

Stanford University School of Medicine

INTRODUCTION

Earlier experiments have shown changes in evoked cortical responses with dc surface polarization (1-4, 8, 12, 14). The purpose of this paper is: 1) to show the intracortical profile of the dc field that is produced by a series of steady currents applied to a point in the cortex, 2) to show the changes in the epileptiform discharge at various levels in the cortex during deep dc polarization, and 3) to discuss one way in which these changes may be interpreted.

EXPERIMENTAL PROCEDURE

The suprasylvian gyrus of cats anesthetized with pentobarbital was exposed. The dc field was produced in the exposed cortex by applying constant current between cortical and mouth electrodes that were connected to a controllable current source. The cortical polarizing electrode was a broken, glass pipette filled with a conducting agar gel (tip diameter about 150 microns). A silver chlorided pellet imbedded in a Ringer's-soaked sponge in the mouth served as the return electrode. The polarizing electrode entered the gyrus at an angle of 30 degrees so the tip could be placed near the tract of an intracortical recording electrode. Positive dc fields and currents were produced when the polarizing electrode was positive with respect to the mouth electrode (anodal polarization); negative fields and currents were produced by reversing the polarity of the supply current (cathodal polarization).

The recording electrode was a glass pipette (about 15 micron

tip diameter) filled with 4M NaCl and mounted on a micrometer drive
assembly. The reference electrode was a second silver chlorided
pellet located in the mouth. The tract of the recording electrode
was normal to the gyrus surface. Signals from the recording pipette
were lead to DC and RC amplifiers and a magnetic tape, FM recording
system (RC band width .1Hz to 1kHz).

Spontaneous interictal epileptiform discharges, with a nomi-
nal seven second interval between discharges, were produced by the
momentary application of penicillin (500,000 units per cc) to the
pia at the site of the pipette recording electrode (9). Recordings
during epileptiform discharge activity were made beginning at a
depth of 2250 microns and proceeding to the cortical surface in 250
micron steps. Currents of 5, 10, 20, 40 and 60 uA of both polarities
were used at selected depths to produce the dc fields.

RESULTS

The profile of the cortical dc field produced by the polarizing
electrode is shown in Fig. 1 for various amounts of polarizing
current. Reversed polarity of the current source produced a negative field and negative currents. The dc field is similar to that
produced by charges moving radially with respect to the electrode
tip in a conducting medium wherein the electrical field decreases
with the square of the distance away from the point of current
application (10). The rate of fall off of the recorded electric
field in the cortex depends upon the relative position of the
polarizing electrode and the recording electrode.

Changes in the epileptiform discharge at various depths in
the cortex during polarization are shown in Fig. 2. The center
column shows the amplitude of the epileptiform discharge at various
depths in the cortex with no field applied (9). The columns to the
left and right of center show the changes that are produced as a
negative and a positive field respectively are applied at a depth

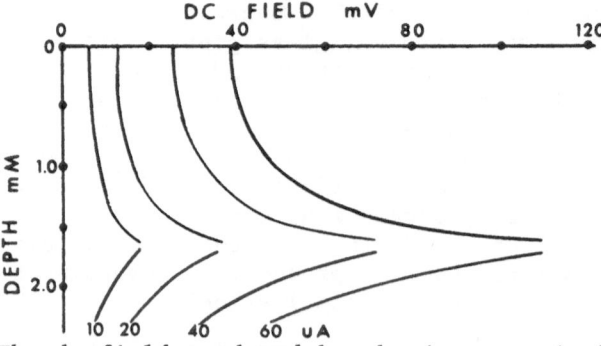

Figure 1. The dc field produced by the intracortical polarizing
            electrode

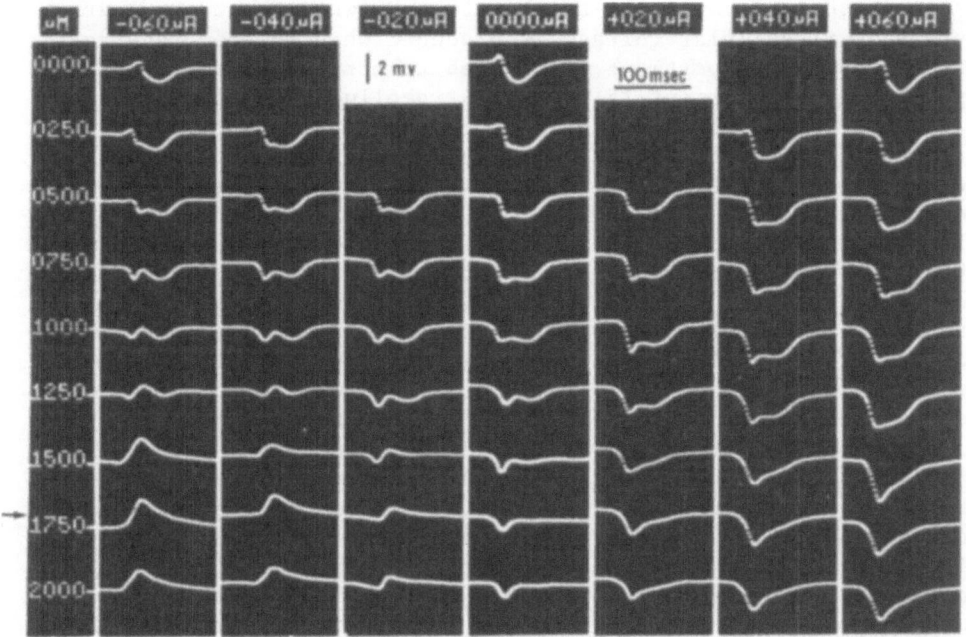

Figure 2. Changes in the epileptiform discharge during deep dc
polarization

indicated by the arrow. The left hand column shows the position
of the recording pipette in the cortex in micrometers. Upward
defections are positive.

The effects of the applied dc field on the epileptiform discharge are consistant with other cortical polarization studies
(1-4, 8, 13, 14). Anodal polarization (positive field) increases
the amplitude of the negative phases while cathodal polarization
inverts the negative phases (13). The greatest changes in the
epileptiform discharges are produced in the vicinity of the greatest
dc gradient while little change is noted at the surface. Quantitative changes in the discharge are shown in Figs. 3A and 3B.
These figures show the magnitude and polarity of the epileptiform
discharge through the cortex at the time of maximum surface positivity (A) and maximum surface negativity (B). The arrows indicate
the position of the polarizing electrode.

DISCUSSION

These data were analyzed in the following way. The population
of dendritic structures that are involved in producing the extracellular synaptic voltages were represented by a two compartment
electrical model of a single dendritic structure (15). This model

was analyzed mathematically to show that between the two compartments an extracellular voltage is produced when the membrane resistance at one compartment is altered to correspond to synaptic
input at the compartment (5). The extracellular voltage produced
at a compartment representing excitatory synaptic activity is
negative with respect to the other (non synaptic) compartment.
The voltage is positive outside of a compartment representing
inhibitory synaptic activity (6). A positive field produced by a
polarizing electrode near the synaptic compartment increases inward
cation flow and decreases outward cation flow. Hence, extracellular
synaptic negativity is increased while extracellular synaptic
positivity is decreased. Opposite effects are produced by the
presence of an applied negative dc field. The synaptic voltage
changes that are produced by the applied fields are compared to
the nonpolarized synaptic voltages. The comparisons are summarized
in Fig. 4A. This figure shows the percent changes in extracellular
synaptic voltages as a function of the applied dc field voltages
for the case where the field is applied at the lower compartment.
Lines 1 and 3 of Fig. 4A represent inhibitory synaptic activity
while lines 2 and 4 represent excitatory synaptic activity. Lines
1 and 2 are for synaptic activity in the deep compartment and lines
3 and 4 are for synaptic activity in the upper compartment. The
direction of the slope of these lines is determined by the direction
of synaptic current and the location of the synaptic activity. The
magnitude of the slope depends on the relative membrane conductance
to the ions carrying the synaptic current, the equilibrium voltages
of these ions, and the membrane voltage.

The data displayed by Figs. 3A and 3B were analyzed by the
method described for the two compartment model. Fig. 3A shows a
positivity of .25mM relative to the synaptic voltage at 1.0mM. With
the upper portion of the .25mM to 1.0mM segment represented by one
compartment and the lower segment by the other, a percent change of
extracellular synaptic voltage as a function of dc voltage between
these depths (Fig. 1) is obtained and shown in Fig. 4B as curve 1.
This curve, when compared with 2 and 3 of Fig. 4A, suggests that inhibitory synaptic activity might exist deeper or both of these
activities may be occurring at the same time. Curve 2 of Fig. 4B
was obtained from Fig. 3B. The slope of curve 2 is similar to 1
and 4 of Fig. 4A and could represent excitatory synaptic activity
near the upper level or inhibitory synaptic activity deeper in the
cortex or both. The greater slope of curve 2 in Fig. 4B could
mean that inhibitory synaptic activity is stronger in the depth
during the surface negative phase whereas the slope of curve 1
(Fig. 4B) would suggest deep excitatory activity during the surface
positive phase (7, 9, 11).

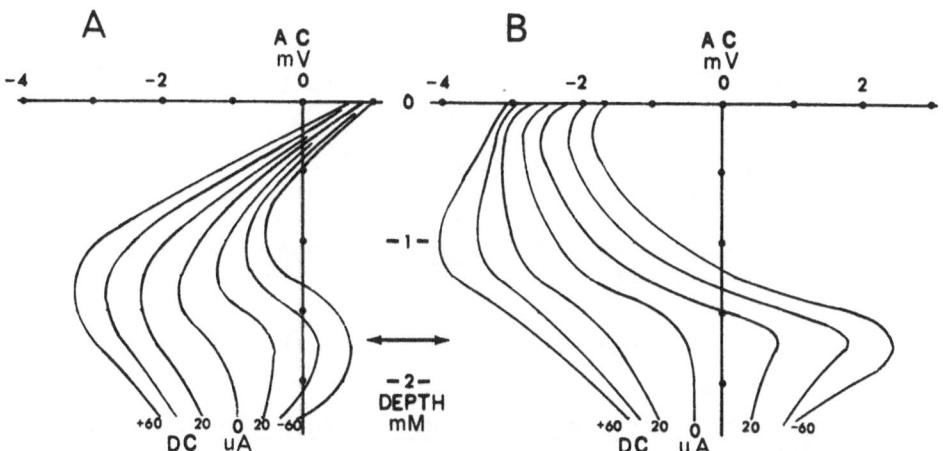

Figure 3.   Laminar analysis of two phases of the epileptiform
            discharge during polarization

CONCLUSION

The dc field produced by a polarizing electrode located in
the cortex is well defined and limited in extent. The effect of
a positive field on the intracortical epileptiform discharge is
to increase extracellular negativity while a negative field reduces
and reverses extracellular negativity in the vicinity of the
polarizing electrode. The results suggest that the combined
techniques of dc polarization and laminar analysis may allow
characterization of the location, sign and temporal sequence of
synaptic activities that generate cortical slow waves.

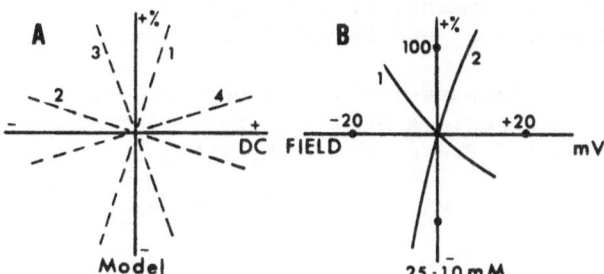

Figure 4.   Percent changes in extracellular synaptic voltages vs.
            the dc field

REFERENCES

1. Bindmann, L.V., Lippold, O.C.J. and Redfearn, J.W.T. J. Physiol., 172:369-382, 1964.

2. Bishop, G.H. and O'Leary, J.L. Electroenceph. clin. Neurophysiol., 2:401-416, 1950.

3. Creutzfeldt, O.D., Fromm, G.H. and Kapp, H. Exp. Neurol., 5:436-452, 1962.

4. Denny, D. and Brookhart, J.M. Electroenceph. clin. Neurophysiol., 14:885-897, 1962.

5. Eccles, J.C. The Physiology of Synapses, Springer-Verlag, 1964.

6. Humphrey, D.R. Electroenceph. clin. Neurophysiol., 25:421-442, 1968.

7. Jasper, H.H. Epilepsia, 2:91-99, 1961.

8. Landau, W.M., Bishop, G.H. and Clare, M.H. J. Neurophysiol., 27:788-813, 1964.

9. Matsumoto, H. and Ajmone-Marsan, C. Exp. Neurol., 9:286-304, 1964.

10. Matveyev, A. Principles of Electrodynamics, Landovitz, translation ed., Reinhold Publishing Co., New York, 1966.

11. Prince, D.A. Electroenceph. clin. Neurophysiol., 23:83-84, 1967.

12. Prince, D.A., Futamachi, K. and Logan, W. Abstracts Intern. Cong. of Neurology and Neurological Surg., Sept., 1969.

13. Purpura, D.P. Comparative physiology of dendrites. In: The Neurosciences: A Study Program, Quarton, et al., (ed.) The Rockefeller Press, New York, 1967.

14. Purpura, D.P. and McMurtry, J.G. J. Neurophysiol., 28:166-185, 1965.

15. Rall, W. Naval Med. Res. Inst. Res. Rept. NM 01 05, 00.01.02: 479-525, 1959.

This work was supported by USPHS Grant Number NB 06477 and Research Fellowship 5 F01 GM 32039.

A PILOT STUDY OF THE EFFECTS OF FUNCTIONAL ELECTRICAL STIMULATION ON THE RECOVERY OF FUNCTION FOLLOWING STROKE IN MAN

P. E. Crago, J. P. Van Der Meulen, and J. B. Reswick

Case Western Reserve University

Cleveland, Ohio, U. S. A.

INTRODUCTION

A preliminary study investigating the possible beneficial effects of functional electrical stimulation (FES) therapy on the degree and speed of recovery of hemiplegic patients has been completed (1).

FES therapy has been used in the past as a treatment for some neuromuscular disorders. Electrical stimulation of denervated muscles to prevent atrophy (2) during re-innervation has been widely investigated and is commonly used clinically. Other effects of stimulation of paralyzed extremities which have been reported are maintenance of joint integrity, prevention of osteoporosis, increased bloodflow, (3) and relief from spasticity and spasms (4). Recently, stimulated muscle has been proposed as an actuator for orthotic systems (5). This results in a combination of both therapeutic stimulation and functional stimulation for movement (6). Active muscular contraction is also of psychological benefit to the patient since he actually can see and feel his paralyzed muscles contracting. FES therapy has been carried out previously on hemiplegic patients by investigators in the Soviet Union (7). Following stimulation, a marked improvement in a patient's voluntary performance (without stimulation) was reported.

It was felt that the effects of electrical stimulation cited above could be of benefit to the paralyzed limbs of a hemiplegic patient and may be helpful in redeveloping voluntary control by the proprioceptive and kinesthetic feedback provided during movements induced by electrical stimulation.

99

METHODS

The study was carried out on four patients: two control patients who received a program of physical therapy as determined by evaluation and two patients who received physical therapy and a program of FES therapy. Treatment and evaluation were performed only on the upper extremity. The non-paralyzed side served as a control for evaluating recovery in each patient, and the group of patients who received only standard physical therapy served as a control for comparison with the other two therapy groups.

When the patient's central nervous system lesion had stabilized, as determined by neurological examination, and if there were no contraindications, for example aphasia or bleeding, the patients were begun on the study. The patients remained hospitalized for three weeks to receive daily treatment and weekly evaluations.

Physical Therapy Program

Patients were given treatment consisting of passive and active range of motion exercise, power building exercise, gait training, and functional activity as indicated by physical therapy evaluation. This evaluation normally involves range of motion measurement of all extremities (8), gross muscle strength assessment of all extremities, and functional evaluation of, for example, transfers and gait. No special attention was given to the muscles being studied, that is, the patients received the same physical therapy they would have received had they not been participating in any study. The patients did not receive occupational therapy.

FES Therapy Program

The FES therapy program used consisted of stimulation for five minutes to each of the following muscle groups of the involved upper extremity; long flexors and extensors of the wrist and fingers, flexors and extensors of the elbow, and the deltoid muscle. A Siemens Neuroton 621 clinical stimulator was used to deliver a 30 ma, current regulated, unidirectional, rectangular pulse of 0.2 msec duration at a frequency of 50 Hz and a surge rate of twelve per minute. A five by six inch dispersive anode was used over the ipsilateral upper trapezius muscle and a two by three inch active cathode was used over the muscles being stimulated. These current levels produced moderate contractions without discomfort to the patients.

Patient Evaluation

The patients were evaluated once each week to determine the degree of recovery and to investigate the time variations of other parameters during recovery. It was desired to minimize the number of parameters as much as possible in order to facilitate the comparison of therapy groups when there is a large number of patients.

To accomplish this reduction, whenever a quantity was measured in more than one muscle, the quantities were combined to yield a single parameter as described below.

Phasic reflexes were assessed on both sides in the pectoralis, biceps, triceps, brachio-radialis and finger flexors by observing the muscular contraction in response to a tap on the tendon with a neurological reflex hammer. The reflex response was graded on a scale from zero to four. The difference was taken between the responses on the affected and unaffected (normal) sides for each muscle tested. Since the difference could be either positive or negative, each one was squared. To make one parameter which would indicate the overall phasic reflex activity, a method of combining numbers which is commonly used in engineering was applied. All of the squared quantities were added, and the square root was taken of this sum. Zero is the normal value.

Tonic reflexes were assessed as the increased resistance to passive stretch on both sides in the finger flexors, supinators and pronators of the wrist, biceps, triceps, and adductors of the shoulder. The grading scale was from zero to four. The same mathematical technique was used as that described for phasic reflexes to make a single parameter to indicate tonic reflex activity. Zero is again the normal value.

Pinch strength is a very functional measure and thus should be a good parameter for measuring the recovery of voluntary strength. A special instrument was designed to measure the isometric pinch strength (1). The pinch meter was positioned between the patient's thumb and his index and middle fingers. The forearm was held securely in a splint attached to a support stand to insure repeatable positioning from one evaluation to the next. The patient was then asked to pinch with as much force as possible. The ratio of the pinch force on the affected side to the pinch force on the unaffected side was used as the parameter.

Hand coordination and function were tested by having the patient attempt to pick up and hold cylinders of various sizes. No restriction was placed on the grasp pattern used. The measured parameter was the number of weeks following the CVA when the patient could first pick up and hold any one cylinder. This test was performed only once each week, hence this number may be larger than the actual time that the patient took to recover the necessary coordination, but not by more than one week.

RESULTS

Since this was a pilot study, the results were only used to show the time variations of recovery parameters. Recovery parameters for a typical patient are shown in Figure 1 as a function of the number of days following the CVA. Phasic reflex activity, tonic reflex activity, and pinch strength are shown. For this

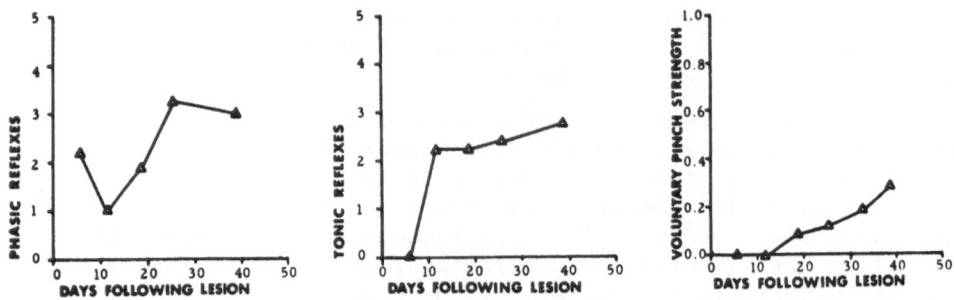

FIGURE 1- PHASIC AND TONIC REFLEX ACTIVITY AND VOLUNTARY PINCH STRENGTH AS A
FUNCTION OF TIME FOLLOWING THE CVA FOR A TYPICAL PATIENT IN THIS STUDY.

patient, the hand coordination and function test movement was first
performed during the fourth week. The reflex activity was still
high in this patient although muscular strength was increasing.

In addition to the parameters discussed above, fatigue and
response to electrical stimulation were measured in the patients.
It was found that these were not useful for evaluating recovery
because they either remained stable, were innaccurate, or were
not indicative of recovery.

CONCLUSIONS

Conclusions cannot be drawn from this study as to which type
of therapy is most effective since the variations among patients
are normally quite large, and the recovery parameters do not show
one group to be markedly different than the other. The investigators feel that the reflex, pinch strength, and hand coordination
and function tests are valid parameters for measuring recovery.

In future studies, it will be necessary to follow patients
until they have reached stabilization. This usually takes less
than six months. However, this would necessitate excessive hos-
pitalization. It would be preferable to have a home therapy program, so that either the patient or a relative would carry out the
procedure. In this case, the patient would be required to come to
the hospital only for evaluation. A new program has been initiated
to carry on the study on a large patient population using a port-
able stimulator designed for this purpose (1).

ACKNOWLEDGEMENTS

The authors would like to express their thanks to Mr. L. Anderson,
Head of the Physical Therapy Department and Mr. C. Taylor and Miss
D. Miller, Staff Physical Therapists, University Hospitals. This
work was supported by Grant Number RD-1814-M from the Social and
Rehabilitation Service, Department of Health, Education, and
Welfare.

1. Crago, P. E., "Evaluation of Therapy Programs and Function Following Stroke in Man," M. S. Thesis, Case Western Reserve University, January, 1970, Engineering Design Center Report No. 4-70-27.

2. Osborne, S. L., "The Retardation of Atrophy in Man by Electrical Stimulation of Muscles," Arch Phys. Med., 1951, 32, pp. 523-528.

3. Wakim, K. G., "Influence of Electrical Stimulation on Circulation in the Stimulated Extremity," Arch. Phys. Med., 1953, 34, pp. 291-295.

4. Levine, M. G., Knott, M., Kabat, H., "Relaxation of Spasticity by Electrical Stimulation of Antagonist Muscles," Arch. of Phys. Med., 1952, 33, pp. 668-673.

5. Crochetiere, W. J., Vodovnik, L., and Reswick, J. B., "Electrical Stimulation of Skeletal Muscle --- A Study of Muscle as an Actuator," Med. and Biol. Eng., 1967, 5, pp. 111-125.

6. Dimitrijevic, M. R., Gracanin, F., Prevec, T., Trontelj, J., "Electronic Control of Paralyzed Extremities: Neurophysiological Considerations," Bio-Medical Engineering, Vol. 3, No. 1., Jan. 1968, pp. 8-19.

7. Aleyev, L. S. and Bunimovich, S. G., "Bioelectric Control of Human Movements as a New Method of Treating Motor Disturbances," 8th International Conference on Medical and Biological Engineering, paper 7-3.

8. American Academy of Orthopedic Surgeons, Joint Motion, Methods of Measuring and Recording, publ. by American Academy of Orthopedic Surgeons, Chicago, Ill., 1965.

INTRAVASCULAR ELECTRICAL THROMBOGENESIS, A PRELIMINARY REPORT

Monty McMinn, B.S.E.E., M.S.E.E.*, C. Roger Youmans, Jr., M.D.*
Wm. Buchholtz, B.S.*, Norman Welford,M.D.**, John R. Derrick,M.D.*

* From the Division of Thoracic and Cardiovascular Surgery
  at the Unviersity of Texas Medical Branch, Galveston, Texas.
** Director of Biomedical Engineering at the University of
  Texas Medical Branch, Galveston, Texas.

INTRODUCTION

Earlier investigators have shown a relation of bio-electrical
phenomena and clot formation. Intravascular thrombosis may be effected by applying an electrical current across a vessel (1,2,3,4,
5,6,7,8), presumedly decreasing streaming and zeta potentials with
resultant disruption of the repulsion forces of similarly charged
blood elements. By positioning transvascular bipolar electrodes in
various vascular compartments and applying constant currents of
varying time and duration, the properties and potentialities of electrical thrombogenesis are presently being evaluated. Accumulated
data at this point permit a small number of conclusions, a few interesting observations, and a great many speculations.

METHOD

Mongrel dogs weighing from 11 to 16 kilograms were anesthetized
with intravenous Nembutal (30 mg. per kilogram body weight). Bipolar
electrodes (either U.S. catheter 5620 or Chardack cardiac catheter
5816) were then positioned in various vascular compartments via the
femoral vein or artery or the carotid artery. A midline abdominal
incision facilitated appropriate positioning of the catheter tip.
A constant current ranging from 0.25 ma. to 3.0 ma. was then applied
through the electrode for periods of fifteen minutes to two hours.
(A multichannel constant current stimulator was designed for this
experiment to permit evaluation of several animals simultaneously.)
At the end of the period of electrical stimulation 50 mg. of

intravenous Heparin was given to prevent further clotting. The vascular compartment containing the electrode tip and the attached thrombus was then removed en bloc.

Any thrombus was then evaluated grossly and placed in 10% Formalin for either microscopic examination or oven-drying and weighing. Coagulation factors were evaluated prior to stimulation and before the final dose of Heparin in each case.

RESULTS

Animals were divided into the following groups: control group, intimal damage group, standard reproducible clot group, anticoagulant group, and experimental and clinical application group.

Control group: In six dogs a catheter was positioned in the inferior vena cava or the infrarenal aorta and left in place for three to four hours. No thrombus or fibrin depostion was noted.

Intimal damage group: Critics of the electrical thrombosis theory suggest that the thrombus results only from associated intimal damage. In order to evaluate this point, the electrode was positioned in compartments not containing intimal linings. In five dogs a woven dacron prosthesis was used to replace the infrarenal aorta. In 15 dogs a silastic arteriovenous shunt chamber was interposed between the femoral artery and vein. A thrombus formed within these chambers in each case with the same time-current product that subsequently produced intravascular clots, demonstrating that electrical thrombogenesis can be independent of intimal damage.

Standard, reproducible clot group: It is well recognized that a standard reproducible thrombus would open new avenues for the field of investigation. In evaluation of this consideration, catheters were positioned in the infrarenal vena cava in 30 days and in the infrarenal aorta in 15 dogs. The charges on the electrode terminals were reversed in several animals to determine the effects of clot formation of reversal of current flow as related to blood flow.

It was found in this group that the size of the thrombus formed was proportional to the rate of current flow and the duration of application of the stimulus. Before thrombus size introduced a factor of hemostasis, there appeared to be a direct relationship between these two factors. Current-time products of 1.0 or greater always resulted in complete vascular occlusion.

Thrombus was always found on the anode, never on the cathode. Changing the direction of current flow as related to the direction of blood flow produced no discernible difference. Microscopic

examination of induced thrombi confirmed a similarity to clots which
form under pathological circumstances, composed of platelets and
fibrin with entrapped red and white blood cells. Greater organization
and fibrin content were present in the larger clots resulting from
the higher time-current products.

It should be noted that although cardiac pacemaker electrodes
were used in this experiment, the impulse from a pulse generator
is bipolar, biphasic and of only 1-2 milliseconds in duration. Indeed, even with constant current stimulation in the 1-3 ma. range,
clots did not form in a ventricular cavity presumedly because a
catheter was "wiped clean" with each heart beat.

Anticoagulant group: The reproducible relationship between the
time-current product and the dry weight of the resultant clot offers
among other possibilities a method for in vivo evaluation of the
effects on intravascular induced thrombosis of various clotting,
anticlotting, and clot lysing agents (e.g. Amicar, Heparin, Dextran,
Urokinase, etc.). Fifteen dogs have been evaluated in this group.
The preliminary results do not permit objective conclusions at this
point. An interesting observation is the fact that recommended dose
of Heparin for human use appears to be in the optimum range for prevention of electrically induced thrombi. Little effect was noted
when the Heparin dose was less than 30 units per kilogram of body
weight. Thrombosis was completely suppressed, however, in four
dogs with Heparin doses of 70 to 80 units per kilogram. Higher
doses of Heparin seemed to offer no additional protection and would
appear to be clinically unwarranted.

Experimental and clinical application group: The ability to
selectively thrombose any major vessel of the body at a given site
simply by fluoroscopy and transvascular catheterization offers certain experimental and clinical advantages. Electrical thrombogenesis
has already been presented as a method for inducing myocardial infarction in the unanesthetized unrestrained animal (9). Additional
uses may include the production of stroke models, renal ischemia
models and others.

Among a number of clinical applications is the concept of
selective organ infarction. Work in our laboratory has indicated
that dispensable organs tend to atrophy and be absorbed after thrombosis of their blood supply. Following this observation, both kidney and splenic infarction with resultant atrophy have been effected
in nine dogs using transvascular electrical thrombogenesis. This
method may be clinically applicable as a form of "electrical organ-
ectomy" without the necessity of surgical intervention. These considerations are only speculative at this point.

SUMMARY

A preliminary report concerning the properties and potentialities of electrical thrombogenesis with a bipolar transvascular
catheter electrode has been presented. It is anticipated that additional studies will confirm that this technique will become increasingly valuable not only in the field of investigation but for
clinical therapeutic application as well.

Bibliography

1. Sawyer, Philip N., and Pater,James W.   "Bio-Electric Phenomena
    as an Etiologic Factor in Intravascular Thrombosis", American
    Journal of Physiology, Vol. 175; 103-107, 1953.

2. Sawyer, Philip N. and Pater, James W.   "Bio-Electric Phenomena
    as an Etiologic FActor in Intravascular Thrombosis", Surgery,
    Vol. 34: 491-500, 1953.

3. Sawyer, Philip N., and Wesolowski, S.A.   "The Electric Current
    of Injured Tissue and Vascular Osslusion", American Journal of
    Surgery, Vol. 153:34-43, 1961.

4. Schwartz, Seymour I., and Richardson, John W.   "Prevention of
    Thrombosis with the Use of a Negative Electric Gurrent", Surgical
    Forum, Vol. 12:46-49, 1961.

5. Sawyer, P.N.; Brattain, W.H.; and Boddy, P.J.,   "Electrochemical
    Precipitation of Human Blood Cells and Its Possible Relation to
    Intravascular Thrombosis", Proc. Nat. Acad. Sci. Usa,Vol.56:
    428-432, 1964.

6. Sawyer, Philip N. and Srinivasan, Supramaniam,   "Studies on the
    Biophysics of Intravascular Thrombosis", American Journal of
    Surgery, Vol. 114:42-60, July 1967.

7. Guest, M. Mason   "Circulatory Effects of Blood Clotting, Fibri-
    nolysis, and Related Hemostatic Processes", Handbook of Physiology,
    Sec. 2, Circ. Vol. III, 1965

8. McMinn, Monty, Derrick, John R., Feldtman, Robert W.   Studies
    and Applications of Electrical Thrombogenesis. Proceedings of
    the Annual Conference on Engineering in Medicine and Biology.
    Vol. 10, 49-B-3, 1963

9. McMinn, Monty, Feldtman, Robert W., Derrick, John R.   A
    Radiofrequency Controlled Myocardial Infarction Animal Model.
    Proceedings of the Southwestern I.E.E.E. Conference and Ex-
    hibition, P. 9D1, April 1968.

A CLINICAL ASSESSMENT OF THE CONDITIONAL EFFECTS OF ELECTROSHOCK

Virginia Johnson, Ed. D.

Private practice
Los Angeles, California

The effects of electroshock (ES) are experienced by individuals in our culture mainly by accident or under medical supervision. Increasingly sophisticated techniques using ES in medicine suggest the need for intensive research into such factors as state-dependent effects, behavioral sequelae, conditioned responses in various neurophysiological systems, and effects upon the memory trace.

In the course of a behavioral therapy program during which subjects were trained in experiential recall (1), memories were recovered from some subjects who had been exposed to ES in varying degrees of intensity, both accidentally and for medical procedures. The purpose here is to report on functional effects which have been recovered in the form of ideomotor memories, and to suggest a possible model within which these effects may be clinically assessed.

The pattern of recall of the shock experience reported by subjects consistently involved the following: (a) involuntary muscular spasms characteristic of those in the original experience; (b) responses simultaneously conditioned at the time of the shock experience, mediated by whatever system or modality was then affected; and (c) altered states of consciousness (ASC) and amnesia, if such effects were present in the original experience.

It is important to note that these recalls were not recovered as single experiences paralleling the time sequence and gross responses of the original behavioral unit. On the contrary, recovery is apparently related to feedback circuits which become associated with discrete elements of the ES experience, on the

111

basis of information already in the nervous system, or in the experiential environment. Another meaningful finding is that as experiential recall of ES approximates the total prior experience, conditional symptoms associated with that experience diminish or disappear. This occurs irrespective of whether such symptoms are mental or physical.

Thus it seems safe to assume that, however encoded by neural information processing for memory storage, the stimulus of ES in its functional aspects is subject to the basic principles of learning and conditioning theory, and the information processing model (2)(3). The relevant question then becomes that of investigating the patterns of response and later behavioral effects of ES conditioning. It is suggested that in this sense ES is a non-specific stimulus which is mediated primarily by the neuromuscular system. Hence the expectancy would be that memory coding would tend predominantly to reflect motor responses, which our subjects consistently precipitate. When the shock effect is sufficient to cause an ASC with its characteristic amnesia effect, then coding involves that state, and the state-dependent effects.

As Strehler points out, "The nature of the patterned input to a system is by itself not the relevant parameter," the key element being rather "the relationship of input patterns to possible alternative responses."(4) This process is characterized by coded signals which form the basis for pattern recognition. In these terms the hypothesis suggested here is that the primary patterns laid down as a response to ES are (a) neuromuscular and (b) altered states of consciousness. Secondary feedback systems are state-dependent upon these patterns.

If tissue areas are destroyed, the problem differs to the degree that coding mechanisms may be rendered incapable of function. The hypothesis suggested here, however, is that any ASC, the stimuli available to the organism during it, and the subsequent state-dependent amnesia are all part of the continuous memory experience. The fact that the information may not be accessible to conscious memory does not negate the possibility of a memory trace which may have a conditional effect on subsequent responses and behavior. The findings from our subjects indicate that the amnesia effect reflects impairment in processes of conscious recall, but not a destruction of the engram.

A confusion exists in much of the literature, both clinical and experimental, because of the assumption that, in the absence of cognitive memory in humans or behavioral responses in animals, no memory trace exists. This is an assumption for which there is as yet no scientific proof. "Not remembering," or amnesia states, are not necessarily equivalent to the absence of engram, only to difficulty with retrieval, since more recovery of preceding events is possible than has been supposed.(5) From the finding that not only was the ES experience itself retrievable in human subjects, but that with its recall the amnesic effect for prior experiences was apparently set aside, the suggestion seems warranted that post-shock amnesia is a function of ASC induced by the shock, but possibly not of the shock itself.

In a review of ECT theories, Miller states that "despite this vast literature and the passage of over 30 years of experimental opportunity, no predominant or convincing rationale for the use of electroconvulsive therapy (ECT) has emerged,"(6) and the evidence is inconclusive with respect to specific effects. In spite of this fact ECT continues to be used on an empirical basis, and, while the effects are highly controversial, there is general consensus that ECT does result in disturbances of memory and confusional states.(7) Most commonly amnesia is experienced both for the shock experience and in varying degrees for prior events.

As Zinkin and Miller have pointed out, behavioral retention depends on "how much of the original trace manages to survive the ECS,"(8) and have questioned whether ECS-produced retrograde amnesia is permanent. Hertz and Peeke have suggested that "re-exposure to the experimental situation is a necessary condition for recovery."(9) McGaugh and Alpern state that the amnesic effect following ES is due to the current, and does not depend upon convulsive movements, and suggest that "subsequent attempts to understand the basis of the amnesic effect of such shocks should focus on the neurophysiological and biochemical effects of electroshock stimulation."(10)

Since EST/ECT are controlled medical procedures, stimulus input and the conditions under which it is administered are relatively constant, and it is proposed that in considering ES procedures prior shock experiences and their state-dependent effects are important variables involving the neuromuscular system and the consciousness continuum. Since most ECT or EST undoubtedly results in ASC which is characterized by amnesia, it acts as a reinforcement of prior ASC, whether ES was the prior stimulus or not. To the extent that the neuromusculature has been conditioned in prior experiences, with ES, a pattern of reinforcement would tend to be established.

SUMMARY

The hypothesis is proposed that ES, however administered or experienced, forms a memory trace coded as a neuromuscular response, and/or as ASC, accompanied by the state-dependent responses in the systems affected by the original experience. Manifestly, however, contextual variables which can become conditional stimuli differ widely, whether intra-organismic or in the external environment.

Ideomotor recall of a shock experience and its accompanying ASC makes available the recall of experiences blocked by the amnesia effect. This would suggest the possible hypothesis that post-ES amnesia is a function of concomitant ASC and not of the shock itself, since recovery of "forgotten" material tends to be elicited after experiential recall of ASC from whatever cause.

If in terms of the information processing model, coding of ES experiences in the nervous system is related to neuromuscular patterns and the continuum of consciousness, then research directed at more intensive study of these two factors may prove clinically useful in the assessment of the functional effects of ES experiences.

REFERENCES

1. Johnson, Virginia. A technique for obtaining experiential recall in behavior therapy. In Richard Rubin and Cyril Franks, Advances in behavior therapy 1968. Symposium presented at the Assoc. for the Advancement of Behavioral Therapies, San Francisco, August 1968.

2. Gaddum, J. The neurological basis of learning. Perspectives in Biol. and Med., Summer 1965, 8, 436-474.

3. John, E. R. Mechanisms of memory. New York: Academic, 1967.

4. Strehler, Bernard L. Information handling in the nervous system: an analogy to molecular-genetic coder-decoder mechanisms. Perspectives in Biol. and Med., Summer 1969, 584-612.

5. Johnson, Virginia. Auditory components of neonatal experience: a preliminary report. Proceedings of the Symposium on Biomedical Engineering, 1, 1966, Milwaukee, Wisconsin, 189-192.

6. Miller, Edgar. Psychological theories of ECT: a review. Inter. J. Psychiat., Feb. 1968, 5(2), 154.

7. Williams, Moyra. Memory disorders associated with electro-convulsive therapy. Amnesia. London: Butterworths, 1966.

8. Zinkin, Sheila, and Miller, A. J. Comment on permanence of retrograde amnesia produced by electroconvulsive shock. Science, June 9, 1967, 156.

9. Herz, Michael J., and Peeke, Harman V. S. Permanence of retro-grade amnesia produced by electroconvulsive shock. _Science_, June 9, 1967, 156.

10. McGaugh, James L., and Alpern, Herbert P. Effects of electro-shock on memory: amnesia without convulsions. _Science_, March 11, 1966, 152, 666.

EFFECT OF ETHANOL ON SOMATOSENSORY EVOKED POTENTIALS

S.A.E. Rosenthal, D.E.Dallmann, F.P. Goldstein,

A. Sances, Jr., and S. J. Larson

Marquette School of Medicine, and Wood VA Hospital,

Milwaukee, Wisconsin

INTRODUCTION

Some studies have been conducted to determine the acute effect of ethanol on the electroencephalographic potential.[1,4,5] However, little is known concerning the central or peripheral nervous system site of action of ethanol.[3,4,5] This experiment was designed to obtain somatosensory evoked potentials from the peripheral nerve, cauda equina, medial lemniscus (ML), ventralis posterior lateralis (VPL), and somatosensory cortex (SSC) of the macaque monkey under the influence of acute and chronic alcohol ingestion.

METHODS AND MATERIALS

For ethanol injection eight adult stumptail macaque monkeys were implanted with Delrin gastric cannulas in the region of greatest curvature of the stomach. For evoked potential recording bipolar nichrome electrodes were chronically implanted in ML, VPL, and SSC. Spinal evoked potentials were recorded from silver half millimeter diameter, spherical tip electrodes placed in the cauda equina through a spinal needle. Evoked potentials from sciatic nerve stimulation were recorded daily for one month to establish controls.

To minimize variation the stimulus electrodes were mounted in a fitted plastic boot which was placed in the same region for each experiment. Each animal was given four grams per kilogram of a 40% by weight solution of ethanol daily through the gastric cannula for a duration of three months. For the first two weeks blood alcohol and evoked potentials were recorded in each animal every half hour

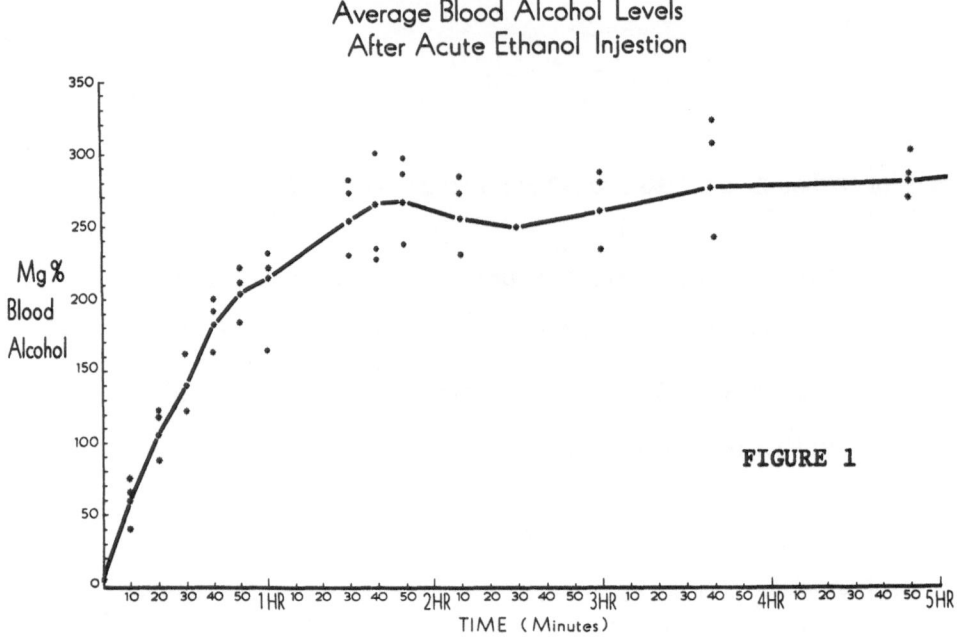

Average Blood Alcohol Levels
After Acute Ethanol Injestion

FIGURE 1

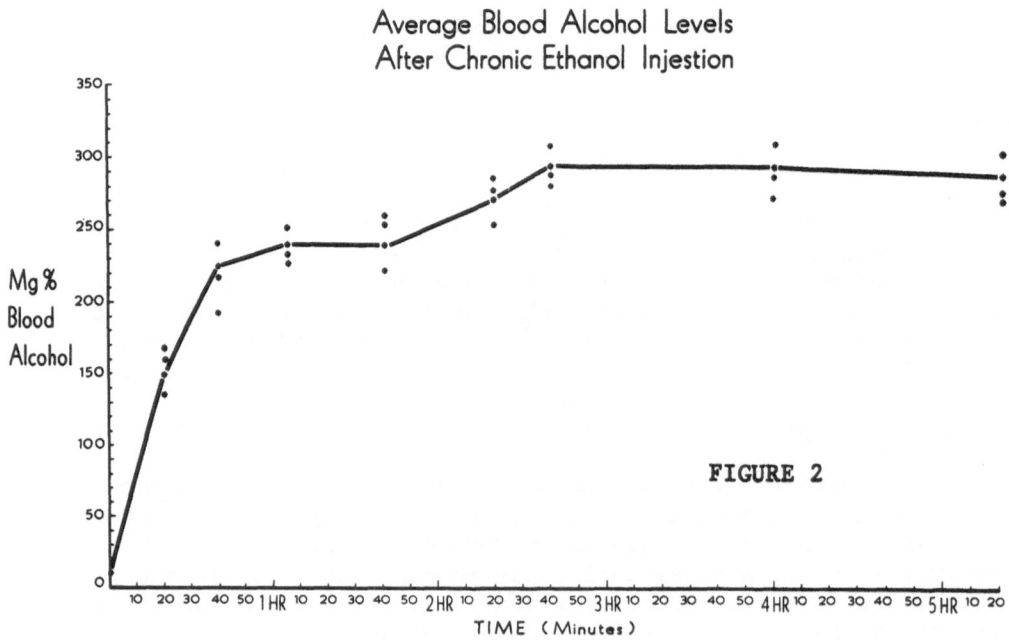

Average Blood Alcohol Levels
After Chronic Ethanol Injestion

FIGURE 2

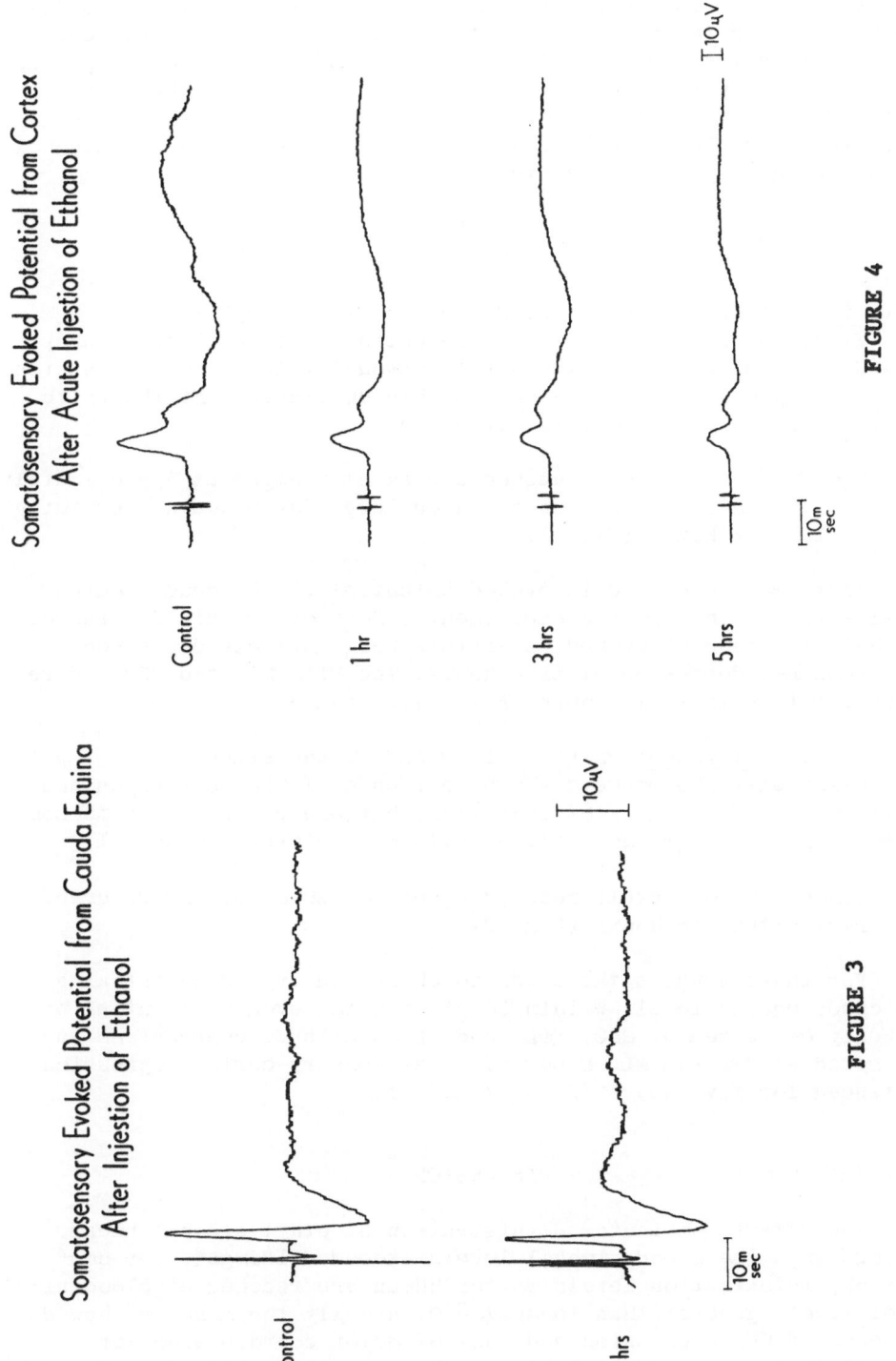

Somatosensory Evoked Potential from Cortex
After Acute Injestion of Ethanol

Control

1 hr

3 hrs

5 hrs

FIGURE 4

Somatosensory Evoked Potential from Cauda Equina
After Injestion of Ethanol

Control

2 hrs

FIGURE 3

for eight hours. Two recordings were made before administration of
the ethanol and the records were taken during the remaining seven
hour period after the injection was made. Additional recordings and
determinations were taken at 12 and 18 hour intervals. Recordings
were obtained over a three month period. After two weeks evoked
potentials and blood alcohols were obtained biweekly. Blood alcohol
was determined by the enzymatic method.[6]

RESULTS

Acute Ethanol Ingestion (First 2 weeks of the study)
     Within 30 minutes of gastric injection of ethanol the monkeys
showed clinical evidence of central nervous system (CNS) depression,
ataxia, progressing to somnolence within 90 minutes. At the fifth
hour, the animals were clinically normal.

     Blood alcohol levels reached a peak of 260mgs% at approximately
100 minutes and maintained an elevated level for 5 hours, returning
to normal by 18 hours (Fig. 1).

     There was no change in evoked potential at the cauda equina
level (Fig. 3) through the experiments. However, within 30 minutes
the SSC, VPL, and ML evoked potentials were markedly depressed.
They remained depressed at five hours. The VPL, ML, and SSC and recovered within 12 to 18 hours (Figs. 4,5, and 6).

Chronic Ethanol Ingestion (2 to 12 weeks of the study)
     Clinically the monkeys showed evidence of CNS depression and
ataxia within 30 minutes of injection, but did not progress to somnolence. By the third hour all animals had returned to normal.

     Blood alcohol levels reached a peak of 290mgs%, and returned
to normal within 18 hours (Fig. 2).

     For these studies there was no change in evoked potential at
the cauda equina level. Within 30 minutes the evoked potential was
markedly depressed at SSC, VPL, and ML. Continued depressions was
not noted at VPL and ML. However, somatosensory cortex depression
continued for five hours (Figs. 7 and 8).

DISCUSSION

     The effects of acute administration of ethyl alcohol became
obvious when the blood alcohol levels exceeded 139mgs%. For comparison, intoxication levels in the human are reached at blood alcohol levels greater than 150mgs%.[2] Clinically the monkeys showed
evidence of CNS depression and loss of motor coordination for
ethanol doses of 4.0g/kg. Acute alcohol administration depressed the

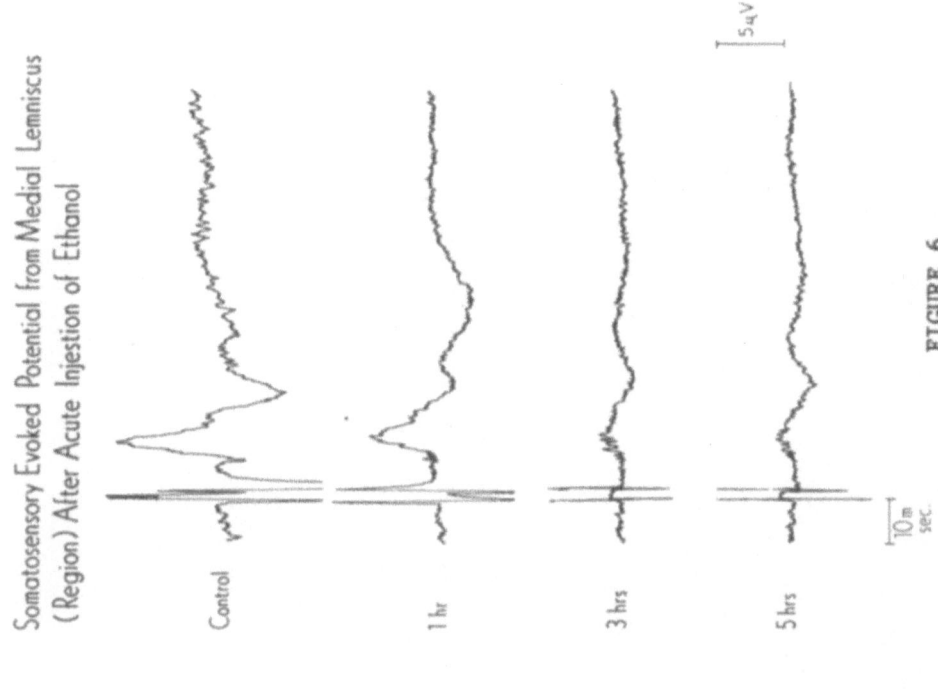

Somatosensory Evoked Potential from Medial Lemniscus (Region) After Acute Injection of Ethanol

FIGURE 6

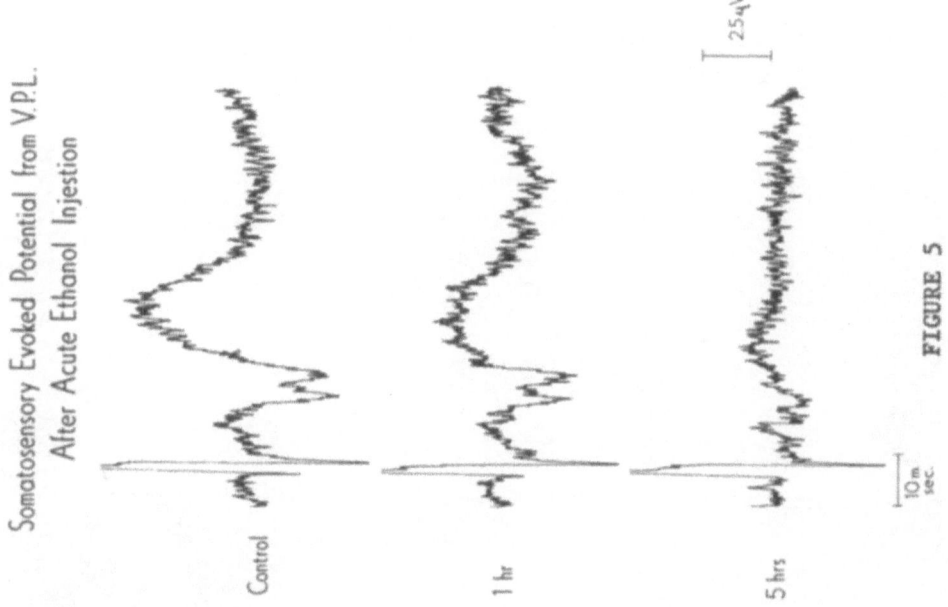

Somatosensory Evoked Potential from V.P.L. After Acute Ethanol Injection

FIGURE 5

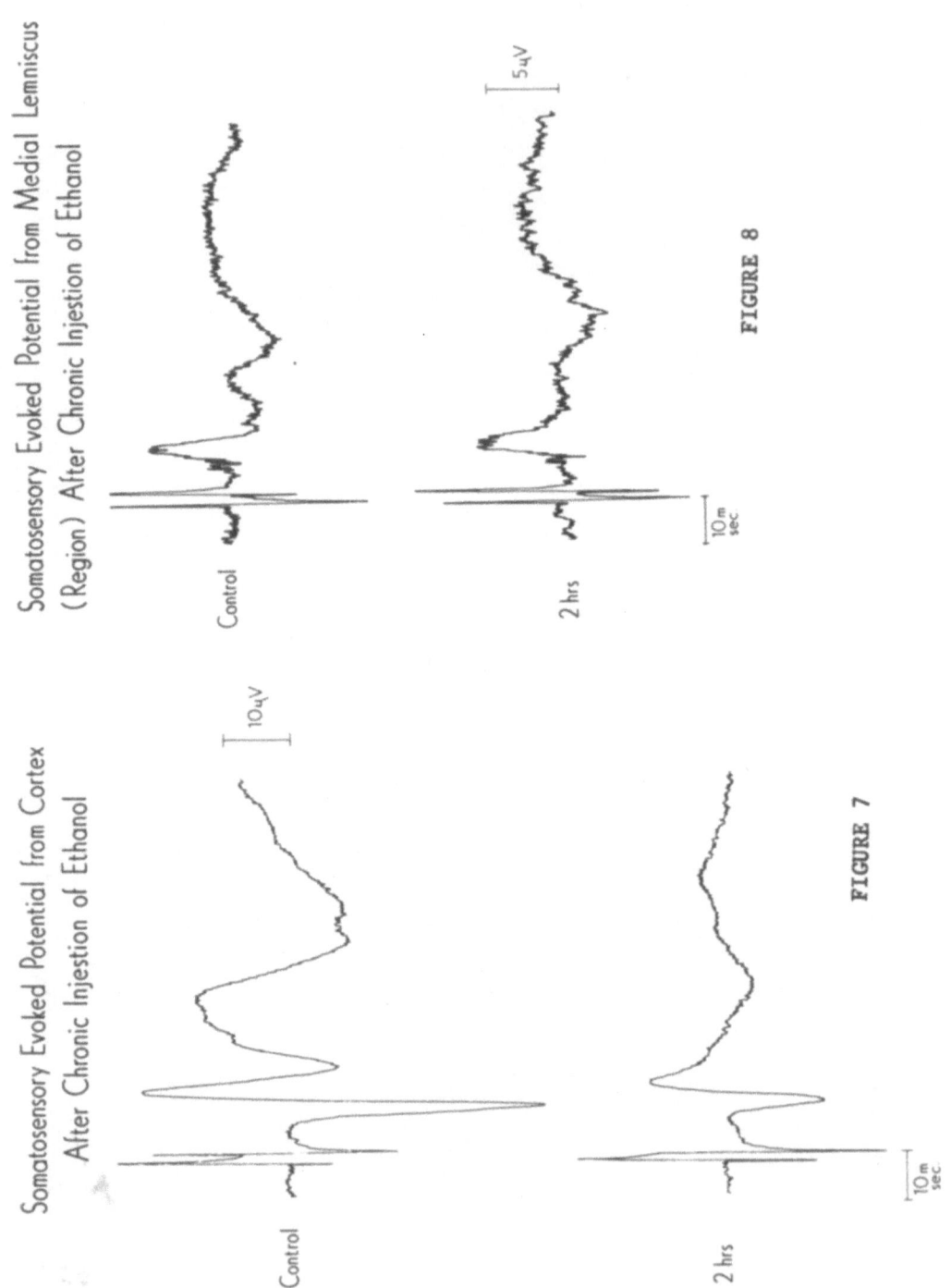

Somatosensory Evoked Potential from Medial Lemniscus (Region) After Chronic Injestion of Ethanol

FIGURE 8

Somatosensory Evoked Potential from Cortex After Chronic Injestion of Ethanol

FIGURE 7

evoked potential at all levels of the CNS. Acutely, depression was more evident in the region of VPL than at SSC suggesting a greater subcortical effect. In contrast, for the chronic studies the cortical depression remained constant, however, the responses from ML and VPL were minimally altered indicating a more profound cortical effect. For these studies the peripheral evoked potentials were not altered in the acute or chronic cases.

The data in the acute state suggests that the ethanol may effect the first synapse. However, accommodation may occur at the cortical level.[4]

However, for the chronic studies inhibition appears to occur above the thalamic level with accommodation at the lower levels.

REFERENCES

1. Hogans, A.F., Moreno, O.M., and Brodie, D.A.: Effects of ethyl alcohol on EEG and avoidance behavior of chronic electrode monkeys. Am. J. Physiol. 201(3):434-436, 1961.
2. Jetter, W.W.: The diagnosis of acute alcoholic intoxication by a correlation of clinical and chemical findings. Am. J. Med. Sci. 196:475, 1938.
3. Victor, M., and Adams, R.D.: The neuropathology of experimental vitamin B6 deficiency in monkeys. Amer. J. Clin. Nutr. 4(4):346-353, 1956.
4. Mirsky, I.A., Piker, P., Rosenbaum,M., and Lederer, H.: "Adaptation" of the central nervous system to varying concentrations of alcohol in the blood. Quart. J. Stud. Alcohol, 2:35-45, 1941.
5. Horsey, W.J., and Akert, K.: The influence of ethyl alcohol on the spontaneous electrical activity of the cerebral cortex and subcortical structures of the cat. Quart. J. Stud. Alcohol, 14:363-377, 1953.
6. Bonnichsew, R.K., and Theorell, H.: An enzymatic method for the determination of ethanol. The Scan. J. Clin. Lab. Invest. 3:58, 1951.

A COMPARATIVE STUDY OF EVOKED UNIT AND POPULATION RETINAL POTENTIALS DURING THE APPLICATION OF DIFFUSE ELECTRICAL CURRENTS

E. J. Zuperku, A. Sances, Jr., and S. J. Larson

Marquette School of Medicine, Marquette University and

Wood VA Hospital, Milwaukee, Wisconsin

INTRODUCTION

Earlier studies have demonstrated that population evoked potentials in the visual[1], auditory[2], and somatosensory[3] systems have been suppressed by the transcranial application of diffuse electrical currents. The greatest reduction in evoked potentials occurred in areas of high synaptic density such as in cortex and retina[4]. Electronmicroscopy studies[5] have shown that the statistical distribution of the number of vesicles per synaptic ending is markedly altered by the electric currents. However, it is still uncertain how the reduction in evoked potentials comes about.

In this paper, the effects of diffuse electrical current upon evoked unit potentials recorded from retinal ganglion fibers will be discussed in relation to the evoked population response recorded from the same fibers.

METHODS AND MATERIALS

The data was recorded from squirrel monkeys. (saimiri sciureus) before, during, and following the current application. A positively biased, unidirectional, rectangular current of 70 Hertz (Hz) and 3.0 milliseconds (ms) duration was applied from the inion to the nasion. A current level of 2.5 milliamperes (ma) direct current (dc) and 2.5 ma average rectangular (ar) produced unresponsiveness in these animals. Diffuse photic stimulation with a rectangular intensity pattern was used.

Population evoked potentials were obtained from nichrome

125

bipolar electrodes chronically implanted in the optic tract. The
electroretinogram (ERG) was recorded with a Burian-Allen B75
contact lens electrode. The current artifact was suppressed by
previously described balancing and averaging techniques.[6]

For unit potential studies, an insulated 1.0 millimeter (mm)
diameter tube was stereotaxically lowered through a trephine hole
in the calvaria to within 5 mm of the optic tract. The lower tip
of the tube was free of insulation and was used for the system
reference point. A micropipette filled with 3 molar KCl was then
hydraulically lowered through this tube into the optic tract.
These animals were prepared and maintained on 20 mg/kg of Nembutal.
The unit potentials were recorded with a Bio Medical Electronics
microelectrode preamplifier. The current artifact was suppressed
by a band limited cancellation network.[7] The data was then stored
on magnetic tape for analysis.

RESULTS

A reduction, with increased electroanesthesia (EA) current,
of the ERG and the evoked population response in the optic tract
was seen (Fig. 1A and 1B). The effects of rectangular current
upon extracellular unit potentials recorded from an "on" type
retinal ganglion fiber were obtained (Fig. 2A). Similar results
were also obtained for "off" type fibers. The current produced
marked alterations in the difference between the mean firing rate
during the excitatory phase of photic stimulation and the non-
excitatory phase of photic stimulation. For direct current levels
of 0.5-2.5 ma dc, the difference was somewhat increased. With
the addition of rectangular current the difference decreased. At
current levels of 2.5 ma dc and 2.0-2.5 ma ar, the difference in
mean firing rate was markedly reduced, and the unit was driven
for short intervals at rates of 70 Hz by the rectangular EA pulse.
After 0.5-2.0 minutes at this current level the unit usually became silent. Five minutes after the discontinuation of the current,
the photically driven unit activity returned to near control levels.
These results are also illustrated by the post stimulus time histogram (PSTH) of Fig. 2B. With the increase of rectangular current,
the PSTH shows the degradation of the photic stimulus pattern into
the background firing. At 2.5 ma dc and 2.5 ma ar the photic
pattern is unrecognizable.

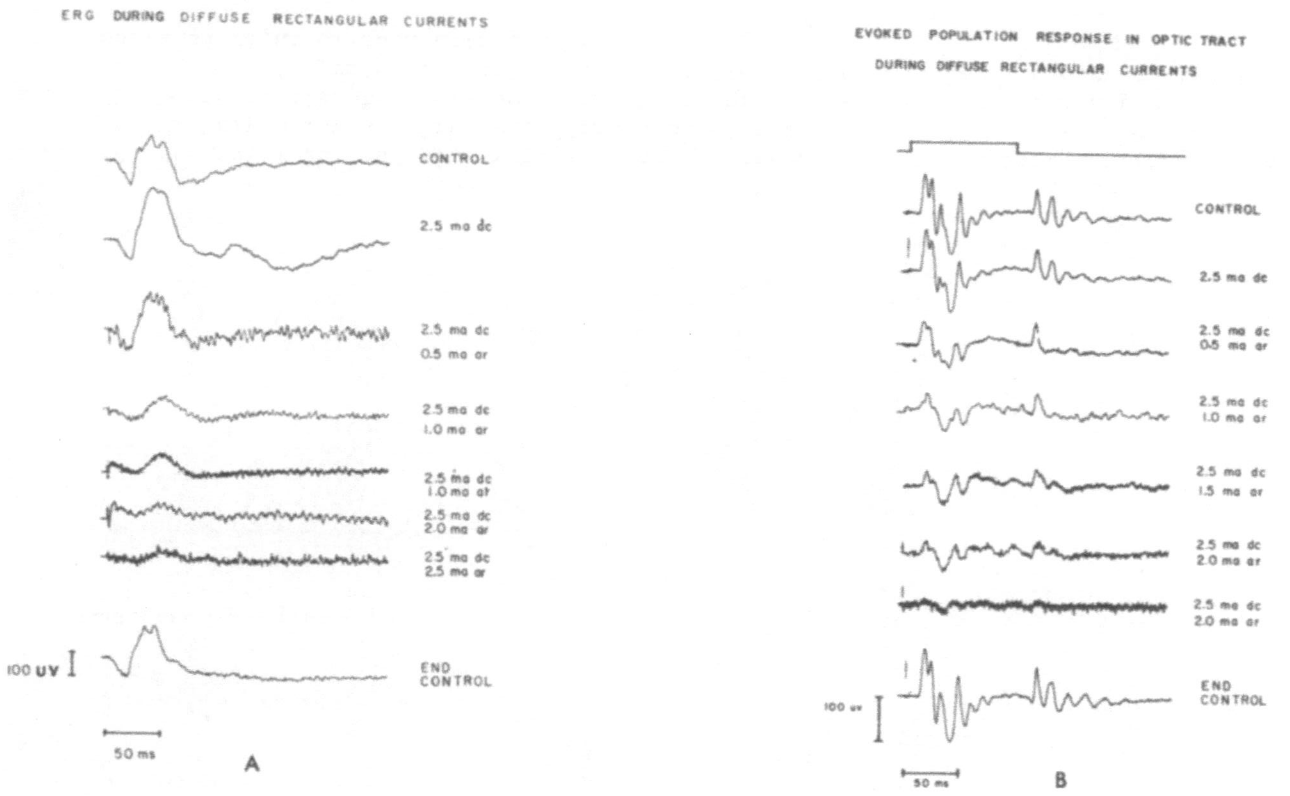

Figure 1: (A) ERG response to strobe flash. (B) Evoked population potential in optic tract. Top trace is photic intensity pattern. The "on" and "off" response is present. (A and B) Each trace an average of 200 responses. The applied current level is shown at the right of each data set. Time and voltage scales pertain to all responses in the respective sets.

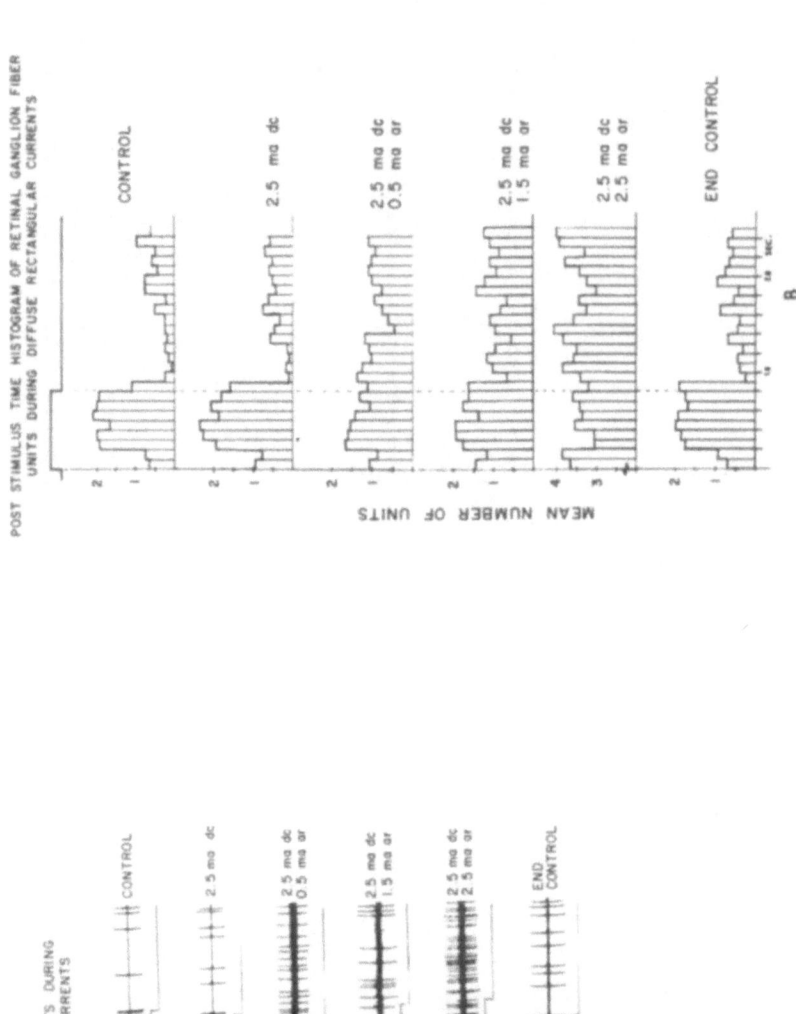

Figure 2: (A) Record of an "on" type retinal ganglion fiber. Light intensity pattern is shown immediately below each record. Time and voltage scale applies to all records. (B) Post stimulus time histogram of the fiber in (A). Twenty records per PSTH. Top trace is photic intensity pattern. Bin width is 100 ms. Applied current level is shown at right of each data set.

DISCUSSION

The preliminary studies show that the diffuse electrical currents alter or mask the unit firing pattern in retinal ganglion
fibers elicited by photic stimulation.

Others have shown alterations in unit potential activity under
the influence of externally applied electrical fields.[7,8,9] An
increase or decrease in unit activity with the application of
direct current is probably a function of the orientation of the
field, the geometry of the neuron, and the cytoarchitecture.
Theoretical considerations based on the electrical properties of
membrane and the geometry of the neuron[10] suggest that external
electric fields cause regional depolarization and hyperpolariza-
tion in the same cell, thus altering the excitability and firing
behavior. Since the electrical currents cause the unit firing
patterns to become non-synchronous with respect to the photic
stimulus, the average response is suppressed. Similarly the difference in firing rate of the units during the excitatory phase of
the stimulus (light on for an "on" fiber, light off for an "off"
fiber) and the non-excitatory phase of the stimulus is not dis-
cernable. Therefore the information content carried by the visual
response to the photic stimulus is masked. Similarly normal
neuronal information carried by the visual networks would be masked
or modified by the EA currents since the coding and decoding
processing in the visual system would probably be markedly affected.
Thus the observed reduction in evoked population responses may not
necessarily result from a lack of neuroelectric activity, but rather
be due to an alteration in the activity.

REFERENCES

1. Zuperku, E. J., Larson, S. J., and Sances, A.,Jr.: Electrical
   anesthesia and the visual pathways, in Electrotherapeutic Sleep
   and Electroanesthesia, F. M. Wageneder and St. Schuy, eds.
   Excerpta Medica Foundation, 1967, pp. 89-94.
2. Larson, S. J., and Sances, A.,Jr.: Physiological effects of
   electrical anesthesia. Surgery 64:281-287, 1968.
3. Sances, A.,Jr., and Larson, S. J.: Physiological mechanisms
   related to electroanesthesia, in Depressed Metabolism, X. J.
   Musacchia and J. F. Saunders, eds., American Elsevier Pub.Co.,
   New York, N. Y., 1969, pp. 39-66.
4. Zuperku, E. J., Sances, A.,Jr., Larson, S. J., and Wilson, A.S.:
   The influence of electroanesthesia on the visual pathways.
   Abstracts of the Conference on the Effects of Diffuse Electrical
   Currents on Physiological Mechanisms with Application to Electroanesthesia and Electrosleep (Milwaukee, Wisconsin)4:36, 1967.
5. Siegesmund, K. A., Sances, A.,Jr., and Larson, S. J.: Effects
   of electroanesthesia on synaptic ultrastructure. J.Neurol.Sci.
   9:89-96, 1969.

6. Sances, A., Jr., and Larson, S. J.: Cortical and subcortical
    bio-potential recording during electroanesthesia. Med. Electron.
    Biol. Engin. 4:201-204, 1966.
7. Toleikis, J R , Dallmann, D. E., Larson, S. J , Sances, A.,Jr.:
    Effects of electroanesthesia upon cerebral unit potentials.
    Proceedings 2nd International Symposium on Electrosleep and
    Electroanesthesia, Graz, Austria, Sept. 1969.
8. Purpura, D. P.: Effects of applied extracellular and trans-
    membrane currents on different varieties of neurons in mammalian
    brain. Abstracts of the Conference on the Effects of Diffuse
    Electrical Currents on Physiological Mechanisms with Application
    to Electroanesthesia and Electrosleep (Milwaukee, Wisconsin)
    4:1, 1967.
9. Creutzfeldt, O. D., Fromm, G. H., and Kapp, H.: Influence of
    transcortical dc currents on cortical neuronal activity.
    Exptl. Neurol. 5:436-452, 1962.
10. Sances, A., Jr., Larson, S. J , and Hause, L. L.: A neuron
    model for electroanesthesia. Proceedings 2nd International
    Symposium on Electrosleep and Electroanesthesia, Graz, Austria,
    Sept. 1969.

SECTION 3

ELECTROSLEEP AND ELECTRO-ANESTHESIA

THE BASIC REST - ACTIVITY CYCLE   (Abstract Paper)

Nathaniel Kleitman, Ph. D.

Professor of Physiology Emeritus, University of Chicago

The basic rest-activity cycle (BRAC) is a physiological period-
icity in the functioning of the nervous system of homoiotheraml
animals. First detected as a variation in the concomitants of sleep
in infants, it involves an alternation of EEG patterns, with a high-
voltage slow activity (HVSA) during the rest phase and a low-voltage
fast activity (LCFA) during the activity phase. The latter is accompanied by certain somatic and visceral manifestations-changes in
heart rate and in respiration movements (REMs), and relaxation of
other muscle groups. In human infants, the BRAC is completed in 55-
60 minutes, progressively increasing to 85-90 minutes in adult man.

The BRACs seem to be of the same mean duration in "good" sleep-
ers, as compared to "poor" ones. Similarly, with the tremendous
increase in the total LVFA sleep time during "recovery" nights,
following multi-night LVFA sleep curtailment, the number of BRACs
per night's sleep remains unchanged. These findings suggest that
the BRAC is a fundamental variation in the activity of the nervous
system.

The round-the-clock operation of the BRAC can be seen in the
distribution of interfeeding periods of neonates, kept on a "self-
demand" feeding schedule: usually a whole number of BRACs--three in
the daytime and four at night. Conditions of living often mask or
interfere with the regularity of the BRAC in adults, but observations of subjects unders controlled conditions support the view that
the BRAC is operative during wakefulness, as well as in sleep.

Does the BRAC manifest itself in electrosleep? Is it preserved
in electroanesthesia and in drug-induced general anesthesia? By
monitoring physiological variables it may be possible to answer
these questions.

133

BASAL FOREBRAIN STRUCTURES AND THE ELECTRICAL INDUCTION OF SLEEP

(Abstract Paper)

Carmine D. Clemente, Ph. D.

Department of Anatomy and the Brain Research Institute,

UCLA School of Medicine, Los Angeles, California

For the last ten years research has been performed at UCLA which has helped elucidate an early finding made in our laboratory, namely that electrical stimulation in the pre-optic basal forebrain of cats and monkeys induces the electroencephalographic and behavioral manifestations of sleep. Experimentation has revealed that the basal forebrain is part of a forebrain inhibitory system which is capable of suppressing widely throughout the body both somatic and visceral mechanisms.

a). It has been shown that electrical stimulation within the basal forebrain including the orbital gyrus will suppress ongoing behavior, induce the behavioral manifestations of the initial stages of sleep and induce synchronization of the EEG.

b). When basal forebrain stimulation is paired with an indifferent tone, conditioning occurs to the extent that after a number of trials the tone alone will induce cortical spindles and sleep behavior.

c). Other studies have shown that if lesions are made within this system there result marked alterations in sleep patterns and generalized behavioral hyperactivity.

d). Electrical stimulation of the basal forebrain and orbital gyrus will markedly inhibit monosynaptic and polysynaptic reflexes in the brain stem and at the cervical and lumbosacral segmental levels of the spinal cord.

e). Stimulation of the orbital gyrus will induce inhibitory post-synaptic potentials recorded within motoneurones of the final

135

common path. Further these hyperpolarizing potentials will change
to depolarizing potentials upon the intracellular infusion of
chloride.

These somatic influences induced by stimulation of the forebrain inhibitory system have been shown to act through inhibitory
centers in the bulbar reticular formation. Behaviorally, reflexively
and intracellularly this system appears to act generally in opposi-
tion to the ascending reticular activation system of the brain stem
core.

Aided by a grant from the U.S.P.H.S.   (MH   10083)

A PRELIMINARY STUDY OF THE USE OF

ELECTROSLEEP THERAPY IN CLINICAL PSYCHIATRY

Ronald R. Koegler, M. D., Shelby M. Hicks, M. D.

Leonard Rogers, M. D., and James H. Barger, M. D.

Olive View Hospital, Olive View, California

It might seem curious that we are reporting on a "preliminary" study of a therapy which has been in constant clinical use for 23 years. Over 500 articles about electrosleep have been published in the Russian literature, describing the treatment of many thousands of patients. However, as Iwanovsky and Dodge pointed out[2] when they reviewed Russian reports, "statistical, experimental, and clinical data were lacking, as was information on the use of controls or efforts to rule out the possiblity of a 'placebo' effect of electrosleep."

Since 1964 interest in electrosleep has increased in Western European countries, particularly Austria and West Germany. Investigators have undertaken sophisticated research and have produced evidence which indicates that electrosleep may be an effective therapy.

Nevertheless, interest and research have lagged in the United States. Clinical psychiatry in this country has preferred psychotherapy over organic treatments; this partially explains the lack of interest in electrosleep. American clinicians view electrosleep with especial skepticism also because the treatment is based on the use of low-amperage electrical current. Electrical machines have been associated with medical quackery in the United States, and this has established an anti-electrical "set" in the minds of the medical profession.

We therefore felt that it was necessary to very carefully "begin at the beginning" with our evaluation of electrosleep therapy. That meant we must first do a preliminary study to help us plan the direction of our more definitive research.

PROCEDURE

1. We encouraged the referral of a wide variety of patients
to our electrosleep project (see Table). The only requirement for
the first group of patients was that they had a sleep problem (Patients #1 through #11)). Although reports in the Russian literature
claimed success with many kinds of psychiatric, psychosomatic, and
medical problems, we suspected that electrosleep was not a panacea
and we wanted to establish preliminary criteria for future patient
selection.

2. To limit the number of variables we settled on a standard
application technique. Using the electrosone 50 transistorized
machine we kept a constant pulse width of 0.5 millisecond, a constant
pulse rate of 40 pulses/second, and a constant time of approximately
one hour. The amplitude and bias (DC voltage) were raised (alternately)
to secure a mild tactile sensation in the eyeball(s) at an initial
current of 0.5 milliamperes (this was done for each treatment, requir-
ing one to two minutes). Electrodes (copper-impregnated pads) soaked
in normal saline were placed over the closed eyes and over the mastoid
processes, held in place by a white mask; a black mask was then placed
over the patient's eyes to secure complete darkness. Treatments were
given five times per week, Monday through Friday, for three weeks
(a total of 15 treatments).

3. The emphasis was on intense observation of a fairly small
number of patients. Therefore all treatments were given completely
by one or more of the authors.

4. Supine, sitting, and standing blood pressures and pulse
rates were taken and recorded just prior to treatment and immediately
afterward.

5. At the conclusion of a treatment the patient was asked to
describe and draw any visual patterns experienced, using colored
pencils.

6. We used a series of check-list ratings to evaluate improvement, plus pre- and post-treatment MMPI tests. Separate check-lists
were filled out by the patient, a friend or relative of patient,
treatment psychiatrist, and patient's therapist (if patient was also
in psychiatric treatment). The check-lists were used in a previous
clinical study.(3) In addition, patients completed a self-rating
depression scale (4) before and after treatment.

RESULTS AND DISCUSSION

Statistical analysis was inadvisable because the wide variety
of patients left us with very few in each sub-category. Therefore

patient	diagnosis	age	hospital or outpatient	sex	number of treatments	insomnia, years	insomnia, severity (*)	change in sleep symptoms (**)	change in other major symptoms (**)	comments
1	Depressive neurosis	32	H	M	10	½	+3	+ +	+ +	(a)
2	Depressive neurosis	61	H	F	15	1½	+3	+ +	+ +	
3	Depressive psychosis	65	H	M	15	1	+3	+ +	+ +	
4	Anxiety neurosis	43	H	F	15	3	+3	+ +	+ +	
5	Anxiety neurosis	24	O	M	15	5	+2	+	+	
6	Obsessive compulsive n.	47	O	M	15	31	+4	+?	+?	
7	Passive-agg. personality	28	O	M	7	4	+1	+	0	(b)
8	Schizoid personality	35	O	M	15	2	+2	+ +	+	
9	Heroin addiction	19	O	M	10	6	+3	+ +	0	(c)
10	Borderline schizophrenia	40	O	M	15	3	+2	+	+	
11	Borderline schizophrenia	37	O	F	15	2	+2	+ +	+ +	
12	Hysterical seizures	35	O	F	(d)	0	0		+	(d)
13	Schizoid personality	17	O	M	(d)	0	0		+	(d)
14	Schizoid personality	20	O	M	(d)	0	0		0	(d)
15	Paranoid schizophrenia	19	O	M	(d)	0	0		+	(d)
16	Chronic schizophrenia	20	O	M	(d)	0	0		+	(d)

TABLE.   Results of Electrosleep Therapy

(*)Severity on scale from "0" (no problem) to "+4" (most severe)
(**)"0" (no change); "+" (improvement); "++" (marked improvement)
(a)Stopped treatment to return to home out of state
(b)Job change. Claimed he was unable to come to clinic
(c)Dropped out
(d)Electro-psychotherapy: electrosleep once per week (or less)
     during psychotherapeutic interview

we cannot make any definitive statements about the value of electrosleep therapy. We did gain impressions which stimulated the form-
ulation of research questions.

Impression #1. Electrosleep therapy seemed to cause improvement
in sleep patterns for almost all patients with moderate or severe
sleep disturbances.

Question: Which patients with sleep problems will not respond
readily to this treatment?

We have some clues. For example, we had anticipated a poor
therapeutic result with patient #6. He is a man whose whole life had
revolved around his insomnia for over 10 years; insomnia excused his
social and personal failures. During the 15 treatments he seemed to
be fighting the possiblity of sleep and his improvement was minimal.
Subsequently we have attempted to engage him in psychotherapy (which
he is also resisting) while continuing electrosleep three times weekly.
He has improved (we think), but he refuses to admit that the improvement has occurred.

Patient #7 had milder insomnia than the other subjects. Apparently his sleep symptoms were not severe enough to motivate him
to continue his treatments when the visits to the clinic interfered
with his daily activities. This may indicate that patients who do
not feel a certain minimal level of "uncomfortableness" are poor
candidates for electrosleep (or for any other psychiatric treatment,
for that matter).

We wondered about the effects of electrosleep in schizophrenia.
However, no schizophrenics (except borderline) were referred to us,
probably because schizophrenia is not characteristically accompanied
by sleep disturbances. The most common psychiatric diagnosis in
our patients was depression.

Impression #2. Significant improvement of non-sleep symptoms
often occurred concurrently with the relief of sleep symptoms.

Question: Is this a direct effect, or secondary to improved
sleep patterns?

We assume that sleep problems can cause or (at least) accentuate
other psychological symptoms, and vice-versa. Hopefully future
research will clarify which is the more common sequence.

In at least one instance we are fairly sure about part of this
sequence. Patient #2 was depressed and also suffered from nightmares. Close questioning before treatment revealed that the nightmares made her afraid to fall asleep and were more significant in
her depressive illness than had been realized by her doctors. She

had a tremendous relief from her insomnia and her depressive symptoms
as soon as the nightmares lessened significantly during treatment.
In this instance the relief of the symptom preceded improvement of
the sleep pattern. However, in other patients the reverse seemed to
occur.

The role of nightmares in sleep problems and depression intrigues
us. Fisher reports that the most significant nightmares occur in
stage 4 sleep[1]; we plan investigations (with EEG sleep recordings)
of the effects of electrosleep on nightmares.

Impression #3. Electrosleep appears to result very quickly in
a close dependence of the patient on the operator (when our technique
is used). This is apparently caused by a combination of: (1) the
supine position; (2) total darkness (we used a black sleep mask);
(3) apprehension about electrical treatment; and (4) the presence of
a psychiatrist-operator in the room for at least part of the treatment.

Question: To what extent are psychological factors responsible
for the clinical effects of electrosleep therapy?

With some patients the operator encouraged conversation. With
others the operator merely started the patient's treatment, then
left him alone. We did not observe a correlation between time
spent with the patient and effectiveness, although we suspect that
(as we evaluate more patients) we will observe instances in which the
psychotherapeutic effects are more important than the direct effects.

We did attempt to observe the effect of combining electrosleep
with psychotherapy once-per-week for patients who were at a stand-
still in their psychotherapeutic treatment (electro-psychotherapy).
Patients 12 through 16 are in this category; all except #14 are still
in treatment. Although it is too early for us to come to any conclusions, the preliminary commentaries of the therapist are interesting:

(Patient #13) "I first saw this patient when he had an acute
schizophrenic break a year ago. Since then I have seen him as an
outpatient. He has always been rather "flat", so one day I suggested
the sleep machine. After about ten minutes on it he surprised me by
'turning on.' He relaxed more than ever before and talked more
freely, mostly about himself and his current feelings and concerns.
More open than ever before. Initially I anticipated a nice quiet
hour with very little talk, during which time I could read a magazine. But he became so talkative and 'ready to work' that I put
away the magazine and worked with him. He can't come every day, but
when he does come in, he wants to use the machine and talk. Asked me
where to buy or rent one."

(Patient #15) "I first saw him 1½ years ago in the hospital
when he was acutely psychotic. Then I treated him as an outpatient
until about two months ago when his car broke down and he said he
could no longer come. Two weeks ago he was again admitted as an
inpatient - same picture as before - confused, loose, concrete, in-
appropriate - very tense, very, very nervous, pacing the floor,
chain-lighting cigarettes. He wasn't much better when he was dis-
charged from the hospital. I suggested the machine to him, hoping
to calm him down. He was very reluctant - paranoid, inspected the
machine, asked many questions, but agreed to try it. After 10 or
15 minutes he started to relax much more than I would have thought
possible. Several times he commented on how much he liked the machine;
twice he asked me to increase the current. Very talkative, but with
more cadence and more organized thinking. Dwelt on current feelings
and thoughts in direct, uncluttered way - not nearly so frantically
ambivalent. At one point he remarked, 'Oh, now this next thing I
want to tell you - could you turn up the machine, I'm going to need
it to say it right.' Fear of machine gone - almost friendly towards
it."

Impression #4. There was considerable individual variation in
the visual patterns which patients perceived during their treatments.

Question: Is there any correlation between visual patterns
and patient characteristics or between visual patterns and response
to therapy?

We encouraged patients to draw their perceptions and have collected some interesting pictures. Patients reported visual phenomena varying from "nothing" to vivid multicolor patterns. Moving
linear black and white patterns were most common; other phenomena
included black and white spirals, silver and gold threads, sheet
lightning, and cones with vivid purple centers. Purple was the
most common color after black and white.

We are particularly interested in patients who have multicolored
visions, remembering that color dreaming is supposedly more common
among schizophrenics. Although we see no obvious correlations we
are still collecting and examining these drawings. It is also possible that we will find some relationship between these phenomena and
the frequency of the stimulating electrical pulses, or with pulse
widths.

Miscellaneous impressions (chiefly technical). Sleep did not
occur before the fifth or sixth treatment (except for one patient).
Relaxation and sleep were experienced more readily and completely
in total darkness (with mask). The rhythmic timer was restful to
some, but irritating to others. Current flow almost always increased
rapidly from 50 to 100% during the first 15 minutes, then more and
more slowly over the remainder of the hour, usually reaching a

maximum of 1.0 to 1.5 milliamperes (100-200% increase); probably this increase in current is secondary to decrease in the skin resistance. A transient blurring of vision usually occurred after the electrodes were removed, lasting 30 seconds to 15 minutes; we hypothesize that this was caused by a minimal displacement of fluid in the anterior chamber of the eye secondary to pressure of the eye electrodes. Although systolic blood pressure drops of up to 50 mm. Hg were noted after treatments as well as substantial decrease in pulse rates, we did not observe dizziness or syncope; we need further studies to determine what role electrosleep (rather than the supine position) plays in the lowered blood pressure, and to determine whether electrosleep has any usefulness in the treatment of clinical hypertension.

SUMMING UP

Our "preliminary study" served its purpose. Results encouraged us to feel that electrosleep therapy may have a place in clinical psychiatry, and convinced us that we should undertake definitive studies. We were able to formulate several hypotheses which we will use as the basis for future studies.

REFERENCES

1. Fisher, C., quoted in Frontiers of Clincial Psychiatry. 6:No. 19 (Nov. 15) 1969, Roche Laboratories.

2. Iwanovsky, I., and Dodge, C. H., Electrosleep and electro-anesthesia - theory and clinical experience, Foreign Science Bull. 4:1-64, Feb. 1968.

3. Koegler, R., and Brill, N. Q, Treatment of Psychiatric Outpatients, N. Y., Appleton-Century-Crofts, 1967.

4. Zung, W. A. Self-rating depression scale, Arch. Gen. Psychiat. 12:63 (Jan.) 1965

ELECTRODE POSITION IN ELECTRO-SLEEP

*N.L. Wulfsohn, F.F.A. (SA) and L. Waldron

*Associate Professor Anesthesiology, University of

Texas Medical School, San Antonio

The production of electro-sleep is partly dependent on electrode position, which determines the area of the brain most affected.
Several varieties of electrode locations are employed by investigators for electro-sleep. The commonest used are fronto-occipital,
bitemporal, one electrode over both eyes and the other over both
mastoid processes, or a cross-over pattern where one pair of electrodes from one generator is over the right eye and left mastoid
process, the other, from a second generator, over the left eye and
right mastoid process.

A variety of currents have also been employed with these positions. They include square waves and sine waves. Recently it has
been shown that an electro-encephalograph modulated sine wave using
a fronto-occipital electrode location can induce a state of electrosleep (Wulfsohn, Condict). To help elucidate which are the most
promising areas of the brain to stimulate for electro-sleep production, this current was applied to different areas of the brain.
This paper reviews this study.

METHOD

Five adult cats were stereotaxically implanted extradurally
with ten electrodes in the same positions, five on each side (see
figure 1). The electrodes were teflon-coated silver coated copper
wire. After three days for recovery from the operation, electrosleep was given employing four volleys of six trains of current
(2.5-5 ma) as previously described (Wulfsohn) using a 90 k sine
wave modulated by the animal's own electro-encephalograph (Condict).
Seven electrode positions were used at each electro-sleep experiment,

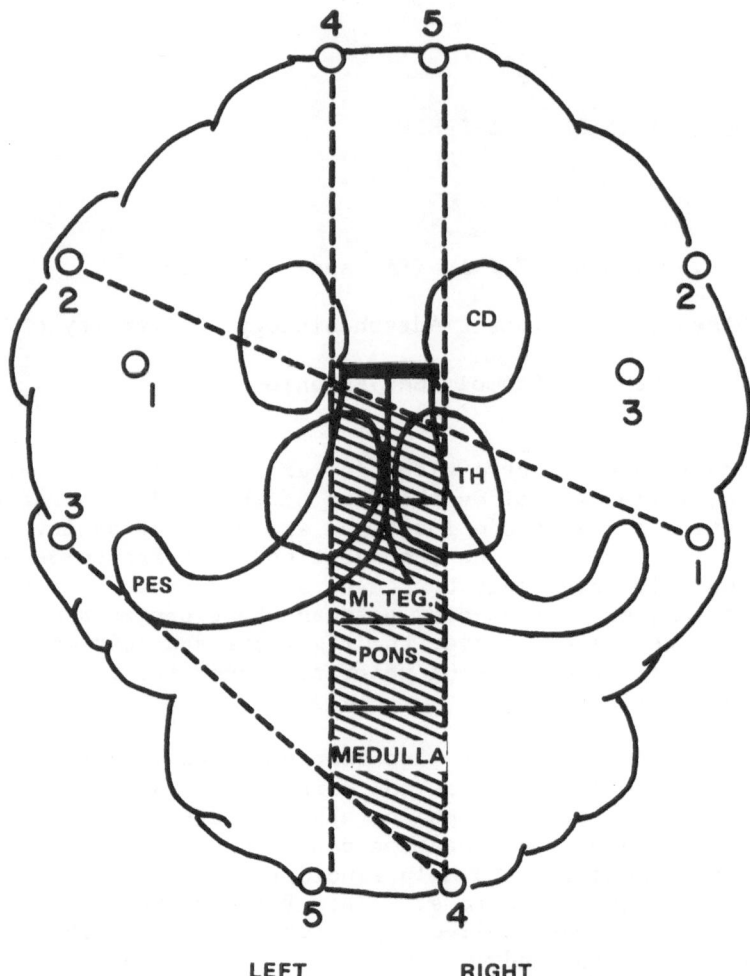

Figure 1

Superior aspect of cat brain with deeper structures superimposed.
Numbers refer to electrode sites. Cross-hatched area is the common
area of stimulation when fronto-occipital and obliquely placed
electrodes are used.

TABLE I

Classification of State of Consciousness of the Cat

SYMBOL	STATE	DESCRIPTION
AA	Awake and alert	Awake and alert, moving about
Q	Quiet	Sitting down, looks about
VO	Very quiet	Sitting down, does not move head about
D	Drowsy	Eyes open and cloes/ or partially closed
VD	Very Drowsy	Eyes closed for short periods Sitting/lying down
LS	Light sleep	Eyes closed (at least 1 minute) opens eyes
DS	Deeper sleep	Nictitating membrane protudes, head resting no response to noise

three for the electro-encephalogram pick-up and four for electro-
stimulation.

The state of consciousness was evaluated according to a chart
showing depth of loss of consciousness (see Table I).

RESULTS

Analysis of the state of consciousness at the end of 3 periods
in each volley i.e. at the end of 6 trains at the end of a constant
current period, and after each rest period revealed that the most
"very drowsy" and "light sleep" periods (45%) occurred in the oblique position (L23,R41) of electrode location. The second best
position (with 31% of these periods) was the fronto-occipital position (L4R5,L5R4) (see Table II).

The same positions, oblique (L23,R41) and fronto-occipital
(L4R5, L5R4), also produced more "very drowsy" and "light sleep"
periods at the end of the fourth volley of trains, 80% for the
former and 50% for the latter. Other positions were less effective
(see Table III).

It was observed that drowsiness and sleep   usually occurred
at the end of the third or fourth volley of trains. It takes 22
minutes to reach the end of the third volley and 28 minutes to
reach the end of the fourth volley.

Controls showed only 18% of "very drowsy-light sleep" periods
occurring at the end of each volley.

DISCUSSION

Trains of current seem to be important in the production of
electro-sleep in cats using an EEG modulated carrier sine wave. A
certain sequence of trains in time and voltage is better than others
(Wulfsohn). Because it takes about 15 to 20 minutes to reach the
sleep stage it is apparent it is a slow process, an "adaptative
phenomenom". The high-fast activity of the awake brain is shifted
into "reverse gear" towards the low-slow activity of the sleeping
brain by this slow adaptation process.

The second best position for electro-sleep L4R5, L5R4 an fronto-
occipital position used in this study is one of the commonest used
by most investigators. The area covered by the third best, an
anterior bitemporal position, is commonly used for electro-anesthesia
and for electro-shock therapy. However, the best position, an oblique one, has not been previously described.

TABLE II

Number of different states of consciousness at end of each volley. Total of 4 volleys in each experiment.

Electrode Position	State of Consciousness							No. of VD-LS-DS Periods	Total	%
	AA	Q	VQ	D	VD	LS	DS			
L35 and R4,1	1	8	5	4	1	1	–	2	20	10
L23 and R4,1	3	2	2	4	4	5	–	9	20	45
L24 and R4,1	6	1	3	5	1	–	–	1	16	6
L23 and R2,1	5	2	4	4	–	1	–	1	16	6
L24 and R2,5	1	4	5	3	–	3	–	3	16	19
L4R5 and L5R4	–	5	3	3	2	1	2	5	16	31
L3R1 and L2R4	1	1	2	9	–	3	–	3	16	19

TABLE III

State of consciousness at the end of the fourth volley of trains.

Electrode Position	Cat No.					Number of VD-LS-DS Periods
	C1	C2	C3	C4	C5	
L3,5 and R4,1	LS	D	Q	VQ	D	1/5 – 20%
L4,R5 and L5R4	VD	DS	D	Q	–	2/4 – 50%
L2,3 and R4,1	VQ	VD	LS	LS	VD	4/5 – 80%
L3,R1 and L2R4	VQ	D	D	LS	–	1/4 – 25%
L2,4 and R4,1	AA	VD	D	VQ	–	1/4 – 25%
L24 and R2,5	AA	VQ	Q	LS	–	1/4 – 25%
L23 and R1,2	Q	D	VQ	D	–	0/4 – 0%

It is interesting to speculate that the structures stimulated by the two best electrode positions for sleep are the structures that lie in their common pathway, i.e. the fornix columns and the medial thalamus.

REFERENCES

Wulfsohn, N., Waldron, L. (1969)   Electrotherapeutic Sleep and
                                   Electro anesthesia. 2nd Inter-
                                   national Congress, Graz, Austria
                                   September.
Condict, E. (1969)   Electrotherapeutic Sleep and Electro anesthesia.
                                   2nd International Congress, Graz,
                                   Austria,   September

Acknowledgement   This research was supported by a grant from the Morrison Trust Foundation, Texas.

ELECTROSLEEP IN MAN BY A COMBINATION OF MAGNETO-INDUCTIVE

AND TRANSTEMPORAL ELECTRIC CURRENTS

D.P. Photiades, R.J. Riggs, S.C. Ayivorh, and J.O. Reynolds

Biophysics Research Unit, Faculty of Pharmacy,

University of Science & Technology, Kumasi, GHANA

INTRODUCTION

The production of electric sleep by low grade pulsed current applied to the human head is very well documented. Although this method is fairly successful therapeutically, it is still in the experimental stage and there is room for improvement. Any method which could potentiate these trans-temporal currents could increase the effectiveness of electric sleep.

With this in mind, we recently showed (8) that pulsed electrostatic fields of 1500 V/cm applied to the human head concomitantly with transtemporal current produced a fairly significant enhancement of electric sleep. Also (9) electro-static fields of 2000 V/cm pulsed five times a second and applied prior to, and continued during two hours of electro-sleep also produced a fairly significant potentiation.

However, the ionization and ozone produced by corona dis-charge, together with the potential danger to human volunteers of more than 2000 V/cm, prompted us to find another means of enhancing cranially applied electrosleep currents.

The initial work of Rentsch (13, 14) on the transmission of currents to the brain with magneto-inductive energy, has shown that this method has certain possibilities in the domain of sleep.

The aim of this preliminary paper is to produce evidence that magnetically induced currents (MIC) concomitant with

153

transtemporal pulsed current (TPC) significantly enhance
sleep.

APPARATUS AND METHOD

TPC was obtained from a standard signal generator which
delivered 100 pulses per second, 1 millisecond in width, and
with an amplitude of 0.5 to 2 milliamperes. This was applied
to the temples of human volunteers via non-polarizable gold
electrodes. The subject was seated in a comfortable, completely insulated chair with arm rests, and with a head rest
which fixed the base of the skull and prevented gross movements of the head when sleep supervened. The body of the
subject was also strapped to the chair to prevent slipping.

MIC was obtained from an air-core solenoid of 2000 turns
of insulated copper wire (18 S.W.G) whose diameter was 30 cm.
A half wave rectified DC source capable of supplying 5000 V
at 0.5 amperes was used to charge up a capacitor, and a heavy
duty contact breaker completed the circuit twice a second.
Silicon rectifiers ensured that oscillations did not occur,
and that only two pulses were delivered every second. The
field strength was approximately 1000 gauss.

Method of Application

With the subject comfortably seated and strapped in the
chair with the base of the skull in the head rest, the gold -
gold chloride electrodes were positioned bitemporally and held
in place by a thin rubber head band. The suspended solenoid
was then lowered until it surrounded the head at the level of

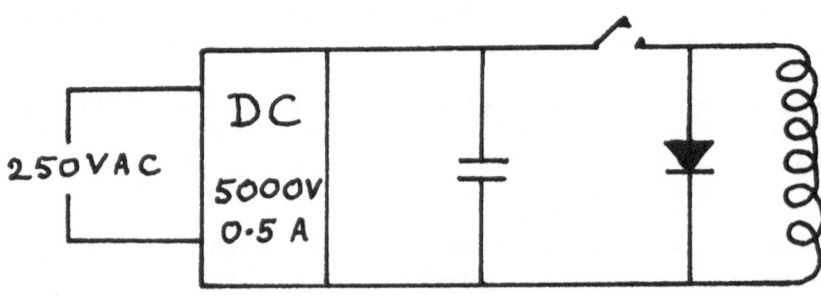

the forehead. The signal generator was then switched on, and
TPC commenced. The subject was allowed 5 to 10 minutes to
adjust the level of current to produce a tingling sensation
at the electrode site. About half an hour after this, the
subject would be asleep. When this happened, the solenoid
surrounding the head would be energized (or it would not be,
depending on the experiment being carried out).

Five human volunteers (three Ghanaians and two expatriate
Caucasians, including one of us--D.P.P.) were each subjected
to TPC for 18 two-hour sessions to produce electric sleep.
The solenoid was energized during 9 of those 18 sessions for
each subject, after electric sleep had supervened. During
the other 9 sessions the solenoid was in position but with
no current flowing. Subjects never knew, and were never told,
if the magnetic field was present or not. Duration and depth
of sleep were the two criteria which counted in these experiments.

Laboratory conditions were maintained as constant as
possible. Each two-hour session was carried out at the same
time of day (before noon) on an empty stomach.

RESULTS

Out of a total of 90 sessions, 24 were failures. This
was either due to the impossibility of TPC to induce sleep,
or due to MIC disturbance.

Depth of sleep was evaluated according to the report of
the subject on awakening, and also according to what appeared
during the experiment. Duration of sleep was evaluated ob-
jectively.

An arbitrarily chosen scale (0, 1, 2, 3) helped in the
final evaluation. On this scale "0" meant no effect or
failure, and "3" meant deep prolonged sleep.

According to this method of evaluation, the results were
as follows: --

	0	1	2	3
TPC (45 sessions)	13	6	22	4
TPC + MIC (45 sessions)	11	5	12	17

This possibly significant enhancement of electric sleep by
magneto-inductive currents merits further study.

                           DISCUSSION

        It is still not exactly known how cranially applied
low grade pulsed current brings about electric sleep, or what
the exact nature of the therapeutic effect is. In electric
sleep, is the effect purely due to the monotonous repetitive
rhythmic nature of the pulses, with tingling sensation at
the electrode site? Is it due to release of chemical substances in the central nervous system which bring about som-
nolescence? Is it due to a combination of the above two
processes, or are other unknown factors involved? The magnetic field from the solenoid induces circular currents in
the brain and movement of charge carriers. It is not known
what effect a field of 1 kilogauss would have on the human
brain, but a number of theories have been advanced recently
to explain the actions of moderately strong fields. Barnothy
(1) states that uncompensated spin motion of the odd electron
imparts a magnetic moment (6) to free radicals (which are
paramagnetic), and this property renders reactions in which
free radicals play a role vulnerable to magnetic fields.
Valentinuzzi (15, 16) implicates the slowing down of rotation-
al diffusion of paramagnetic molecules by externally applied
fields as capable of retarding biochemical reactions.

        Gross (4) states that biomagnetic effects may be due to
a depression of paramagnetic enzyme-substrate activity due
to distortion of bond angle or orbital of such molecules by
a magnetic field, while the possibility that magnetic fields
act as non-specific stressors producing hormone imbalance
should be borne in mind (2).

        Kholodov's experiments (5) have shown that in rabbits
exposed to 800 Oe static fields, there was a significant increase in the number of high amplitude slow waves in the
occipital region, indicating an inhibition of the central
nervous system (CNS). He also showed that lesions of the
diencephalon caused disturbances of conditioned reflexes to
magnetic fields. He therefore concluded that such external
fields exerted a direct influence on the diencephalon as
proved by encephale isole preparations.

        Other behavioral correlates of weak magnetic fields
have recently come to light. Friedman and coworkers (3)
have shown that magnetic fields of about 10 gauss from a
coil, pulsed once every 5 to 10 seconds, significantly increased reaction time in human subjects, although the probable

way in which this occurred in the CNS was not mentioned.

Persinger's work (7) on the open field and shock avoidance behavior in rats exposed prenatally to a 3 to 30 gauss 0.5 cycle rotating static magnetic field showed significant behavioral changes in test rats. He chose prenatal exposure since he argued that formation of the CNS and other organs was rapidly taking place during that period, and therefore they were then most susceptible to external influences. Again he does not mention how these weak magnetic fields act.

In our opinion, the chemical basis of biomagnetism (at molecular level) in the CNS merits further study. We are now determining (10) the optimum magnetic field in the solenoid (using intensities of 2 and 3 kilogauss, and also 500 and 100 gauss), with different repetition rates at the contact breaker. We are also investigating (11) the effect of fronto-occipital low grade pulsed current in combination with magneto-inductive energy in humans, and also testing the urine of subjects (12) for breakdown products of chemicals previously implicated in sleep mechanisms.

ACKNOWLEDGMENTS

We are grateful to the Dean, Faculty of Engineering, for the loan of transformers and rectifiers; also to the Head, Department of Physics, for the loan of measuring instruments.

Our thanks are due to Mr. Leo Miller for his help in the laboratory and also for agreeing to be one of the volunteers, to Dr. M. Valentinuzzi of the Chicago College of Osteopathy and Physiology, Chicago, Illinois, for useful suggestions, and to Mrs. Carol Takyi for typing the manuscript.

REFERENCES

(1) Barnothy, J. M. Biological Effects of Magnetic Fields. Ed. M. F. Barnothy, Plenum Press, New York, 1964, p. 3.

(2) Barnothy, M. F. and Sumegi, I. Abnormalities in organs of mice induced by a magnetic field. Nature, 1969, 221, 270.

(3) Friedman, H., Becker, R. O., and Bachman, C. H. Effect of magnetic fields on reaction time performance. Nature, 1967, 213, 949.

(4) Gross, L. <u>Biological Effects of Magnetic Fields</u>. Ed.
     M. F. Barnothy, Plenum Press, New York, 1964, p. 74.
(5) Kholodov, Y. A. <u>Biological Effects of Magnetic Fields</u>.
     Ed. M. F. Barnothy, Plenum Press, New York, 1964, p. 196.
(6) Mulai, L. N. <u>Biological Effects of Magnetic Fields</u>.
     Ed. M. F. Barnothy, Plenum Press, New York, 1964, p. 33.
(7) Persinger, M. A. Thesis, University of Tennessee,
     Knoxville, Tenn., 1969.
(8) Photiades, D. P. and Ayivorh, S. C. Enhancement of
     electric sleep by electrostatic fields disturbing
     central nervous homeostasis. 5th Internat. Congress
     Cybernetic Medicine, Naples, 1968.
(9) Photiades, D. P. and Ayivorh, S. C. The potentiation of
     electrostatic fields. 2nd Internat. Symp. Electrosleep
     and Electroanesthesia, Graz, Austria, 1969.
(10) Photiades, D. P., Ayivorh, S. C., Riggs, R. J., and
     Reynolds, J. O. Unpublished data.
(11) Photiades, D. P., Ayivorh, S. C., Riggs, R. J., and
     Reynolds, J. O. Work in progress.
(12) Photiades, D. P., Riggs, R. J., Reynolds, J. O., and
     Ayivorh, S. C. Work in progress.
(13) Rentsch, W. The application of stimulating currents
     with magneto inductive energy transmission. Digest 6th
     Internat. Conf. Med. Electr. Biol. Engng., Tokyo, 1965.
(14) Rentsch, W. Magneto-inductive transmission of stimulat-
     ing current pulses to the brain. 1st Internat. Symp.
     Electrotherapeutic Sleep and Electroanesthesia, Graz,
     Austria, 1966.
(15) Valentinuzzi, M. <u>A Theory of Magnetic Growth Inhibition</u>.
     Univ. of Chicago, Ill., 1963.
(16) Valentinuzzi, M. <u>Biological Effects of Magnetic Fields</u>.
     Ed. M. F. Barnothy, Plenum Press, New York, 1964, p. 63.

ELECTROSTIMULATION OF DEEP BRAIN STRUCTURES

* N.L. Wulfsohn, F.F.A.(SA), A. Davis and E. Gelineau

* Associate Professor Anesthesiology, University of
  Texas Medical School, San Antonio

Electro-sleep can be produced in cats by superficial cortical
stimulation using trains of current with an electro-encephalograph
(EEG) modulated sine-wave current (Wulfsohn and Waldron). Where
this current produces its effect is not clear. To help elucidate
this problem this current was used to stimulate some of the depth
structures in the brain already shown to produce sleep by the use
of other currents, such as low frequency square waves or pulses.

METHOD

Four adult cats were stereotaxically implanted with depth
electrodes for stimulation of the following structures - caudate
nucleus, thalamus,fornix columns, pes hippocampi and pre-optic
area. Twelve different areas or combination of areas were stimulated. Each experiment was repeated three times on different oc-
casions with a total of 78 experiments including controls (see
Table I).

Stimulation was carried out with four volleys of trains of an
EEG modulated sine wave current, with 1 to 2 ma., followed by a
constant current period 1 ma for 3 minutes, and a zero current
rest period of 2 minutes.

The effects on the state of consciousness were evaluated on a
scale previously described (Wulfsohn, 1969). Readings were made
after each volley of trains after each constant current period and
each rest period.

TABLE I

Percentage of sleep periods produced by electro-stimulation in various depth areas in 3 experiments in in 3 experiments per cat. Cats used were C-7, C-8, C-10, C-11.

Areas Stimulated Electrode I	Electrode II	% Drowsy-Light-Deeper Sleep Periods			% Light-Deeper Sleep Periods			% Deeper Sleep Periods		
		Cat C-7	Cat C-8	Av. of 6 Expts.	Cat C-7	Cat C-8	Av. of 6 Expts.	Cat C-7	Cat C-8	Av. of 6 Expts.
1. FX L.	FX R.	83	58	70	67	25	46	36	6	21
2. CD L.	CD R.	64	53	59	25	8	17	0	0	0
3. L. CD ant.& R. CD ant.	L. CD post & R. CD post	–	36	36	–	0	0	–	0	0
4. CD L.& FX R.	CD R. & FX R.	56	31	44	33	6	20	8	0	4
5. CD L.	FX R.	50	53	52	29	11	20	6	0	3
6. CD R.	FX L.	50	47	49	19	6	13	0	0	0
7. CD L. & CD R.	FX L. & FX R.	39	36	38	25	3	14	6	0	3
8. FX L. ant. & FX R ant.	FX L. post.& FX R post	–	47	47	–	11	11	–	0	0
9. Control		28	17	23	0	0	0	0	0	0

Electrode I & II Placement	Cat C-10	Cat C-11	Av. of 6 Expts.	Cat C-10	Cat C-11	Av. of 6 Expts.	Cat C-10	Cat C-11	Av. of 6 Expts.
10. Within L. med. thalamus	47	19	33	14	9	12	3	0	1½
11. Within L. fornix	47	28	38	19	13	16	0	0	0
12. Within L. pes hippocampi	44	11	28	31	8	20	3	0	1½
13. Within L. preoptic area	14	0	7	0	0	0	0	0	0
14. Control	8	19	14	0	0	0	0	0	0

Results

The percentage of drowsy and sleep periods following electro-stimulation are shown in Table I. In cats 7 and 8 where large areas are covered by the stimulus the stimulated structures which produce the most "drowsy, light sleep and deep sleep periods" are the columns of the fornix (70%) the caudate nuclei (59%), and the crossed area caudate nucleus to the fornix columns (52%-49%).

In the control there were only 23% of these periods. When considering just light or deep periods and excluding the drowsy periods the best effect is again seen with stimulation of both fornix columns (46%), the next best with a crossed area caudate to fornix (20%). Stimulation of the caudate nuclei alone produces 17% of these periods. The control had none of these periods(0%).

Deeper sleep periods again occured most frequently (21%) in the fornix columns following stimulation.

In both cats 10 and 11 where much more localized areas were stimulated the most "drowsy-light sleep-deeper sleep periods" occurred in the left fornix 38%, the left medial thalamus 33% and the left pes hippocampi 28%. Considering just the production of light and deeper sleep periods the best effect was found in the left pes hippocampi (20%) the left fornix (16%) and the medial thalamus (12%). Deeper sleep periods occurred in the medial thalamus only 1½% of times and the pes hippocampi 1½%.

Discussion: In our experiment the most promising area for production of sleep by electro-stimulation is the fornix column. Wherever the fornix columns were involved with other areas in electro-stimulation sleep was produced to some extent.

Stimulation of the caudate nucleus, pes hippocampi and medial thalamus were good but not quite so good as the fornix columns for sleep production. This is not the same area as found by other workers.

Electrical stimulation of several areas in the brain can produce slow wave or synchronized sleep. Areas that have successfully responded to low frequency stimuli of pulses or square waves are the bilateral basal forebrain area(Sterman and Clemente, 1962) reticular formation of the mesencephalon (Proctor, Knighton and Churchill, 1957: Favale et al.,), of the pons (Magnes et al, 1961), and medulla (Favale et al., 1961); of the head of the nucleus caudatus, mamillary body, nucleus lateral thalamus, midbrain tegmentum fornix (Parmeggiani, 1962); certain cortical surface areas, such as premotor, somesthetic, anterior and posterior suprasylvian and posterior supra-orbital areas (Penaloza-Rojas, Elteman and Olmos, 1964); nucleus of the solitary tract (Magnes, Moruzzi and Pompeiano,

1961): and thalamus (Parmeggiani, 1961-62; Akert et al., 1965), (Akert, Koella, Hess, 1952 and Monnier, 1963).

Desynchronized or paradoxical sleep can also be produced by direct electrical stimulation of the pons (Rossi, 1963; Jouvet, 1965). However, this effect can only be produced when the cat is in a state of fully developed synchronized sleep.

Acknowledgement   This research was supported by a grant from the Morrison Foundation, Texas.

REFERENCES

1.   Akert, K., Koella, W.P., Hess, W.R. (1952) Sleep produced by electrical stimulation of the thalamus. Am. J. Physiol. 168-260.

2.   Akert, K., Bally, C., Schade, J.P. (1965) Sleep Mechanisms. Progress in Brain Research. Vol. 18,11

3.   Favale, E. Loeb, C., Rossi, G., Sacco, G. (1961) EEG synchronization and behavioral signs of sleep following low frequency stimulation of the brain stem reticular formation. Arch. Ital. de Biol., 99, 1-22.

4.   Jouvet, B. (1965) Sleep mechanisms: Progress in Brain Research Vol. 18, 20.

5.   Magnes, J., Moruzzi, G., Pompeiano, O. (1961) Synchronization of the EEG produced by low electrical stimulation of the region of the solitary tract. Arch. Ital. de Biol. 99, 33.

6.   Monnier, M., Hoshi, L., Krupp, P. (1963) Moderating and activation systems in mediocentral thalamus and reticular formation. The Pharmacological Basis of Mental Activity. EEG and Clin. Neurophysiol. Supplement 24., 97.

7.   Parmeggiani, P. L., (1962) Sleep behaviour elicited by electrical stimulation of cortical and subcortical structures in the cat. Helv. Physiol. Acta, 20, 347.

8.   Parmeggiani, P.L., (1961-1962) Pflugers Archiv. fuer die gesamte Physiol. 274,85.

9.   Penaloza-Rojas, J.H., Elterman, M., Olmos,N. (1964) Sleep induced by cortical stimulation. Expt. Neurol., 10, 140.

10. Proctor, L., Knighton,R., Churchill, J. (1957) Variations in consciousness produced by stimulating the reticular formation of the monkey. Neurol., 7,1,193.

11. Rossi, G.F., (1963) Sleep inducing mechanisms in the brain stem. In: the physiological basis of mental activity. Ed. R.N. Peon. EEG and Clin. Neurophysiol. Supplement 24, Elsenier Publishing Co. N.Y.

12. Sterman, M.B., Clemente, C.D. (1962) Forebrain inhibiting mechanisms: sleep patterns induced by basal forebrain stimulation in the behaving cat. Expt. Neurol., 6, 103.

13. Wulfsohn, N., Waldron, L. (1969) The importance of trains of current in electrosleep. 2nd Inter. Congr. Graz, Austria, September.

CLINICAL EXPERIENCES OF ELECTRO-ANAESTHESIA (eA)

+ Brigadier K.R.Rama Rao, FFARCS,FAMS, DA.
++ Major Sitaram Bhat, M.D.,
+++ Major M.S.Rajagopalan, D.A.

In our continuing studies in the fascinating field of electro-Anaesthesia (eA) at Army Hospital Delhi Cantt-10 India since 1962 as an Armed Forces Medical Research Committee Project, we worked extensively on dogs for nearly two years (Rama Rao, 1962; Rama Rao and Sitaram 1962). This enabled us to gain good experience in its technology and establish well defined clinical parameters for eA and safety factors (Rama Rao et al 1966).

This paper deals briefly with our clinical experiences in this field indicating differences in physiological correlates of different placements of electrodes.

THE GENERATORS

The generators used in the study have all been designed and developed by Electronics and Radar Development Establishment (LRDE) Bangalore, India. Apparatus for eA Mk IA for sinusoidal alternating current, Mk II for A.C.pulse with D.C.Bias and Mk III for wide band noise are the three generators in use (Figs 1, 2 & 3).

TYPES OF ELECTRODES

Metal disc electrodes – made of silver coated copper discs of 2.5 cm diameter, wire-mesh electrode made of silver coated phosphor-bronze wire mesh, spectacle electrodes which is a liquid

----

+ Consultant in Anaesthesia, Army Hospital, Delhi Cantt 10 India.
++ Specialist in Anaesthesia, Army Hospital, Delhi Cantt-10 India.
+++ Specialist in Anaesthesia, 153 GH, C/O 56 APO.

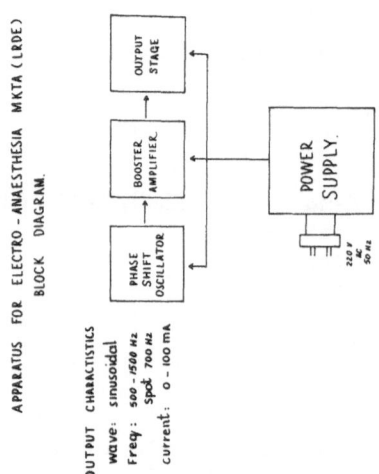

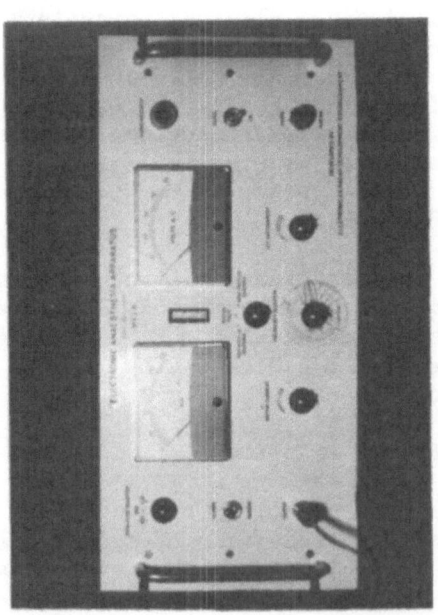

Figure 1

Apparatus for eA Mark I-A (LRDE) with Block Diagram and output characteristics.

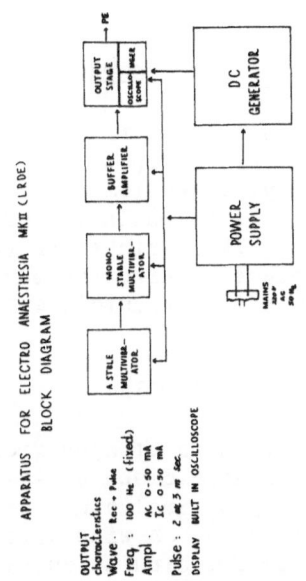

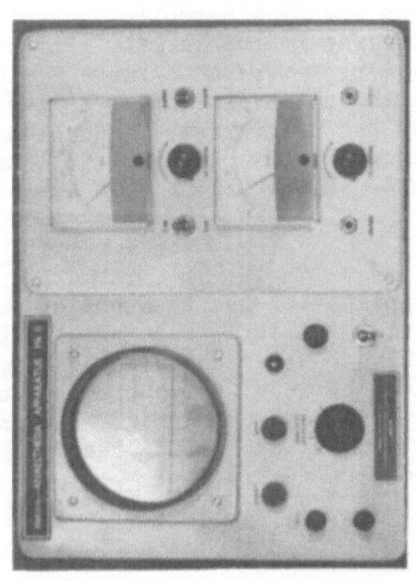

Figure 2

Apparatus for eA Mark II (LRDE) with Block Diagram and output characteristics.

phase electrode incorporating saline soaked gauze in a
"despatch rider" spectacle for surface placements and a stainless
steel needle electrode for subcutaneous placement are the four
types of electrodes in use.

PLACEMENTS OF ELECTRODES

Three have been surface placements being trans-temporal
(bi-temporal), fore-head occiput and eye occiput, and the three
others subcutaneous placements namely, trans-temporal, brow
occiput and trans-parieto-mastoid. The nomenclature of the first
five placements is self explanatory. The sixth trans-parieto-
mastoid placement is one which is placed behind the ear above the
mastoid process obliquely along the parieto-mastoid suture line -
hence this name.

CONDUCT OF ANAESTHESIA

250 patients (210 males and 40 females of ages from 3 to
66 years) without any co-existing systemic disease were selected
for eA for elective and emergency surgery including two almost
moribund patients (Rama Rao et al 1966 b). A wide variety of
surgical procedures ranging from 6 to 200 minutes duration have
been undertaken.

Pre-anaesthetic medication was by a narcotic analgesic agent
like Morphine sulphate or Pethidine hydrochloride. Atropinisation
was done uniformly in all cases by intravenous route 5 minutes
prior to induction. To avoid induction discomfort a sleep dose of
thiopentone was used initially and not repeated. No other para-
medication was resorted to except lissive doses of gallamine
triethiodide as muscle relaxant. Respirations were assisted/
controlled depending on the degree of muscular relaxation required
for the surgery contemplated.

The pattern of current application has been one of relatively
fast input at the rate of about 10 mA in 15 seconds. Both the
'blow in' input, which is likely to cause undesirable cardiovascular
effects or convulsions, or the slow input which produces only a
prolonged period of struggle, have been abandoned. Patients were
interrogated 1 hour, 24 hours and one week after the operation to
assess intra-operative awareness or memory of it

EVALUATION OF OBSERVATIONS

Our evaluation of anaesthetic state has been on clinical
parameters, like reaction to stimuli, muscular phenomena, eye signs,
respirations, pulse and blood pressure, discussed in a previous
paper (Rama Rao et al 1966 a). Depending on these clinical
parameters for onset and progress of anaesthesia, and post operative

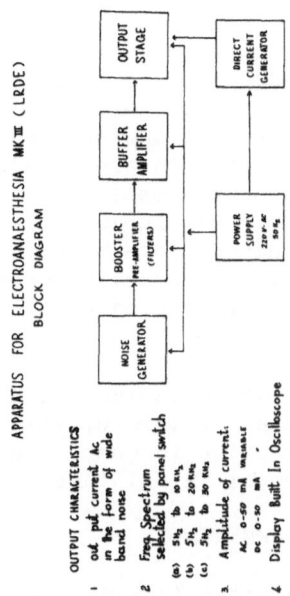

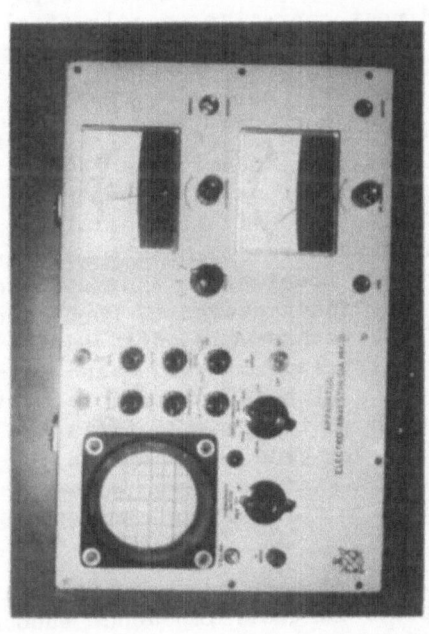

Figure 3

Appartus for eA Mark III (LRDE) with Block Diagram and output characteristics.

observations including intra-operative awarness cases have been
graded as 'successful', 'partially successful' or 'failures'.
Clinical criteria for grading of the results have been as discussed
by Rama Rao and Sitaram Bhat (1969).

In general our observations have been mild to moderate degree
of tachycardia, hypertension, muscular phenomena like twitches,
clonic and tonic spasms, and increased exocrinal activity. The
degree of each of the physiological correlates of electro-
Anaesthesia differed with different types of current and different
placement of electrodes. These in our opinion are considered to
be within the physiological tolerance of the patient. Other
undesirable side effects noted in some cases included local burns
at the site of electrodes (in three cases), some memory of
intra-operative procedures (in five cases) and clinic convulsions
during immediate post operative period (in one case which did not
recur and EEG was normal).

Recovery pattern has been uniformly instantaneous after
switching off the current with immediate return of vital reflexes
but for the residual effect of curarisation in some cases. Nausea
or vomiting has been conspicuously absent in all these cases.

RESULTS – COMMENTS

Our studies have been chiefly directed towards evaluation
of anaesthesia, by clinical parameters, observation of physiological
responses indicated above and post operative sequele of electro-
Anaesthesia with three forms of current, and six placements of
electrodes.

Of the three types of currents used, Sinewave AC has given
consistently sound anaesthetic state. Square wave pulses caused
excessive induction discomfort, marked muscular phenomena, increased
sympathomimetic activity. DC bias necessitates use of surface
liquid phase electrode, which is cumbersome and its make and break
of contact is difficult to avoid. Local burns that were noticed were
all with direct current. As the disadvantages of DC bias outweigh
the possible advantage it offers, we have discontinued the use of
DC bias both with square wave pulse and with wide band noise.
Frequencies of 1200 Hz and 1500 Hz sine AC were found to give better
results than a frequency of 700 Hz which was the frequency used by
Knutson et al (Knutson et al 1956). Wide band noise with a band
width of 5 Hz to 30 KHz without DC bias has given encouraging results
and induction discomfort has been noticed to be less with this type
of current.

Subcutaneous placements of electrodes have considerable
advantages over surface placements in that; impedence encountered
was markedly less, keeping the electrodes in place was not a problem
and make and break of contact was avoided. The only disadvantage is
the inability to use direct current.

We have observed that nearer the electrodes are to the cortex more marked are the muscular phenomena and nearer the electrodes to hypo-thalamic region more marked are the autonomic disturbances, in terms of increased sympatho-mimetic and exocrinal activity. Assuming that thalamus is the 'target' area in the production of anaesthetic stage (Tatsuno et al 1967), and taking into consideration the surface anatomy in Indian subjects, we have tried a placement termed as "trans-parieto-mastoid" placement which aims at maximum current density at posterior part of thalamus including pulvinar. We have observed that cardio-vascular, muscular and exocrinal side-effects have been minimal with this placement. Differences noticed with these six placements have been discussed in a previous paper (Rama Rao and Sitaram Bhat 1969).

Our limited clinical experiences gained so far, have shown us that, electro-anaesthesia is a distinct possibility and there has been no mortality or morbidity directly attributable to the method. The chief disadvantages have been induction discomfort, lack of muscular relaxation and mild to moderate sympathetic over activity.

The most important pre-requisite - apart from the other essential requirements in terms of wave form and frequency - for the production of electro-Anaesthesia is the placement suitable electrodes on the closed skull. Wulfsohn (1966) suggested shifting the then commonly used surface trans-temporal (bitemporal) placement to a slightly higher and posterior position with the external auditory meatus as a land mark.

Our results suggest that 'trans-parieto-mastoid' subcutaneous placement of stainless steel needle electrodes provide the basis for minimal deviations of physiological correlates. These results cannot be claimed as conclusive unless verified and confirmed by others in the field of research.

ACKNOWLEDGEMENTS

We are grateful to the Director General, Armed Forces Medical Services India and the Director of Medical Services (Army) for having given us the opportunity to work in this new and fascinating field and for the permission given to present this paper and participate in the proceedings of this conference.

We thank Dr Norman L. Wulfsohn for providing this unique opportunity to share and exchange our mutual experiences in the field with other workers.

We gratefully acknowledge the pioneering work done by Electronics and Radar Development Establishment Bangalore India, in designing and developing all the generators and accessories used in the study.

REFERENCES

1   Knutson, R.C., Tichy, F.Y., Reitman J.H (1956): The use
     of Electric Current as an Anaesthetic Agent:
     Anaesthesiology 815-825.

2   Rama Rao K.R (1962): Electro-narcosis - Experiments on
     dogs and preliminary observations: Armed Forces Medical
     Journal of India: 1 : 18 ; 317-327.

3   Rama Rao, K.R., and Seetharam (1962): Electro-narcosis:
     Some experimental and clinical aspects: Ind J.Anaesth,
     10: 2: 69-76.

4   Rama Rao, K.R., Rao, L.N., Bhalla S.K (1966 a): Proceedings
     of the first international Symposium on electro-sleep
     and electro-Anaesthesia: Graz Austria. pp.266.

5   Rama Rao, K.R., Rao, L.N., Bhalla S.K (1966 b): Clinical
     signs of electro-Anaesthesia: Proceedings of the first
     International Symposium on electro-sleep and electro-
     Anaesthesia. Graz, Austria, pp. 269.

6   Rama Rao, K.R. and Sitaram Bhat (1969): Electro-Anaesthesia:
     Clinical Experiences with different placements of
     electrodes: under publication.

7   Tatsuno, J., Zouhar, R.L., Smith, R.H., and Cullen S.C
     (1967): Electro-Anaesthesia studies: The Target Area
     for electro-Anaesthesia: Anesth & Analg, 46:432-439.

8   Wulfsohn, N.L (1966): Clinical electro-Anaesthesia:
     Proceedings of the first International Symposium:
     Graz, Austria, pp 276.

ELECTROANESTHESIA STUDIES: SITE OF CURRENT ACTION

Jiro Tatsuno, M.D., Ph. D., Robert H. Smith, M.D.,

Hidehiro Suzuki, M.D., Ph. D., and J. Herbert Andrew, M.D.

Department of Anesthesiology, University of California
Medical Center San Francisco, California

INTRODUCTION

This laboratory is engaged in a continuing inquiry into the
phenomena associated with electroanesthesia. We have felt that we
should have reliable information on the "site" of, or areas critical
to, electroanesthesia production, the best way to get maximal current to that site, and what happens to the area in the presence of
current. This present study was designed to yield information on
the site of electroanesthesia. In 1966 we reported (1) results of
a study which indicated that the diencephalic area was that portion
of the brain which must receive sufficient current if the anesthetic
state were to be produced. The present study was designed to provide a more precise localization of the target area. The basic
idea was to inject a dielectric liquid into the third and lateral
ventricles to ascertain whether this obstacle to current passage,
by changing the current path, would change the current requirements
for electroanesthesia.

When a dielectric obstacle is interposed into a current path
it modifies that path in a predictable manner. The current does
not flow to the dielectric and the abruptly change its course to go
around the obstacle. Instead the current goes around in long smooth
curves which develop some distance from the obstacle. As the current stream splits to flow around the obstacle the diverging streams
leave an "island" of low density current between the streams and
the dielectric material.

In this study this "island" of low current density included
that portion of the thalamus which is lateral to the third ventricle.

173

METHOD

Three adult Rhesus monkeys were the subjects. For each, the electroanesthesia current requirements were ascertained. The current employed was 1,000 cycles per second (cps) sine wave applied via subcutaneous needles in the temporal areas 1 centimeter anterior to the external auditory meatus on both sides. The anesthetic state was considered to be adequate for the test when the animal would tolerate 30 seconds of severe tail-clamping without bodily movements or with-drawal.

After the base line current requirements for electroanesthesia were established, each animal was anesthetized with halothane, and had appropriately placed burr holes made in the skull. 0.7 to 0.9 cc of ethyl iodophenylundecylate (Pantopaque), a dielectric liquid, were injected via #25 needles (through the burr holes) into the third ventricle. Roentgenograms verified that the Pantopaque was in the third and lateral ventricles.

The animals were permitted to waken and were observed for any evidence of brain injury or untoward reaction to Pantopaque.

When awake and responding normally to stimuli, each animal was again subjected to electroanesthesia in the same method as before. The evaluation of the "normal" state comprised:

1. The animal's response to an opportunity to bite anything close to its face.

2. The response to threatening gestures and sounds directed toward the animal's face.

3. The response to offered water.

4. The response to pinching of the extremities.

5. The response to being carried from the operating table to the restraint chair.

6. The evaluation of the strength of the extremities.

A few weeks after the test, x-rays were repeated, and the animal was again subjected to electroanesthesia.

RESULTS

Each animal required at least 28% or more current to produce electroanesthesia after Pantopaque was injected into the third and/

or lateral ventricles (Table I).

Table I

Monkey	Electroanesthesia Base line requirements	After Pantopaque Current requirement	Increase Above Base line	Four Weeks Later
1	22 mA	28 mA	28%	21 mA
2	24 mA	32 mA	33%	24 mA
3	24 mA	35 mA	46%	24 mA

Each animal reacted normally after it wakened from halothane-induced anesthesia with Pantopaque in the third and lateral ventricles.

The second x-rays, taken at least four weeks after testing, revealed that the Pantopaque had drained out of the ventricles. The electroanesthesia requirements of each animal were found, at that time, to be the same as they were in the original base line studies.

DISCUSSION

We believe this study very strongly suggests that the portion of the thalamus lying immediately lateral to the third ventricle is critical to electroanesthesia production. We can not delineate, as yet, the exact dimensions of the area, nor describe any other portions of the thalamus that are critical. The inquiry will continue.

A few points in the study should be emphasized.

The electrode placement is critical. The subcutaneous needle electrodes, placed 1 centimeter anterior to the external auditory meatus, provide greater current density in the thalamus than does any other placement on the sides of the head (2).

Pantopaque is a very good dielectric. Its current-shunting characteristics were studied extensively in our current-flow test-cell. When the third and lateral ventricles had been injected we anticipated that the column of Pantopaque would divert the current around it and away from the brain tissue lying immediately lateral to the third ventricle.

After the Pantopaque had drained out of the ventricles into

176   J. TATSUNO ET AL.

the spinal dural sac the test animals' current requirements were
what they were before the Pantopaque was injected; there was no
permanent change in requirements.

   The dielectric substance in the lateral ventricles did not
deflect the current away from those portions of the thalamus lying
beneath the floor of the lateral ventricles. In fact, the Pantopaque presence may have increased the current density in that area,
since current could not traverse the lateral ventricle filled with
the dielectric liquid, and had to go above and/or below the ventricle.

   The fact that electroanesthesia could be produced by the application of enough additional current suggests that some part of
the diencephalon other that that portion of the thalamus lying along-
side the third ventricle must be involved to some extent in electroanesthesia production. This test only provides evidence concerning
that portion of the thalamus closest to and lateral to the third
ventricle.

                              SUMMARY

   When the third ventricle of a Rhesus monkey is filled with a
dielectric liquid (Pantopaque) the current requirements for the
production of electroanesthesia are increased 28% or more. This
finding suggests that the portion of the thalamus lying lateral to
the wall of the third ventricle is involved in the production of
electroanesthesia.

Acknowledgements:

This research was supported by U.S. Army Contract DA-49-193-MD-2424.
The authors gratefully acknowledge the use of facilities at the
San Francisco General Hospital (San Francisco, California) and the
technical assistance of Peter A. Lindquist, Robert H. Dempsey, and
Mrs. Shelley Frisch.

References

1.   Tatsuno, J., Zouhar, R.L., Smith, R.H., and Cullen, S.C.:
     Electroanesthesia studies: the target area for electroanesthesia.
     Anesth & Analg., 46:432 (1967).

2.   Tatsuno, J., Zouhar, R.L., Smith, R.H., and Cullen, S.C.:
     Electroanesthesia studies: location of electrode sites on the
     skin to provide maximal current density in the thalamus area.
     Anesth & Analg., 46:163 (1967).

accelerometer, 80

acetylcholine, 82

action,
  potential, 58,66
  current, 66
activity, bioelectric, 5

addition, 139

alcohol, 117

amblyopia, 33, 35

amnesia, 112

analgetic effect, 91

anesthesia, electro-165,168
  (see electro-anesthesia)

angle, phase, 41

anxiety, 139

aorta, 106

asthenia, 68

ataxia, 120

atrophy, 99

atropine, 79

auditory,
  system, 125
  tone, 57, 58

behavioral therapy, 111

biological systems, 3½

bioelectrical activity, 5

biomagnetism, 157

bipolar electrodes, 15,51
blind, 38
blood flow, cerebral, 21
  brain, 82

blood pressure, 138, 143

blood vessel, 22, 24, 25,
                105
blurring of vision, 143
brain, 71

brain resistivity, 24

bradycardia, 68

burns, 91
  local, electrodes, 170

cannula, gastric, 117

cardio-vascular, 171

carrier frequency, 30

cat, 59, 145

catecholamines, 81

cauda equina, 120

caudate nucleus 82, 161

cephalography impedance, 21

cerebral blood flow, 21

cerebro-spinal fluid, 18

cholinergic drugs, 79

clamp, tail, 174

clot, 105, 107

color stimulation, 36

colored light, 35

compulsive, 139

conduction, nervous, 9

consciousness, 111, 147, 159

constant, dielectric, 10

convulsions, 170

convulsive shock, 27

cortex somatosensory, 117

cortical potentials, 37

current, 3, 9
  action, 66
  density, 10, 15,
  ionic, 9
  intensity, 90, 91
  levels of, 79
  low density, 173
  magneto-inductive, 153
  measurement, 3
  rectangular, anesthesia, 126
  requirements, 176
  site of action, 173
  shunting, 175
  trains of, 148, 149
  transtemporal, 153,154

DC currents, 71
  fields, 93
  polarization, 93
deaf, 86

densities, 7

density, current, 10, 15, 16, 173

depolarized, 78

depressed, 140

dermographia, 68

detector, magnetic, 10

Deutsch's, probe, 16, 18

dielectric substance, 18, 82,
                      173, 174
  constant, 10
diencephalon, 156

dipole, 24, 25, 36

dog, 28

dopa, 79

dopamine, 79

drowsy, 148, 161

electrical current,
  diffuse, 125

electrical fields, 10, 24, 129
  resistance, 5

electrical thrombogenesis, 105

electro-anesthesia, 15, 27,79,
                    81,126,133,
                    165,173,
                    174,176
  (see also anesthesia)

electro-cardiograph (ECG),
                    28, 31
electro-encephalograph(EEG)
            33, 117, 133, 135,
            145, 159
  changes of, 57,69

electro-encephalograph (EEG)
  modulated sine wave, 159

electro-magnetic field, 3, 4
  measurement, 3
  radiation, 85

electro-phonic hearing, 85

electro-psychotherapy, 141

electro-retinogram, 126

electro-shock, 82, 111, 113,
  effects of, 111

electro-sleep, 79, 133, 135, 137,
        145, 153, 155, 159
  (see also sleep)

electro-static field, 153
  shielding, 9

electro-stimulation, hearing, 85
  denervated muscle, 99, 100

electro-therapy

electrode, 15, 21, 35, 40, 41, 42,
       138
  bipolar, 15, 51, 74, 105
  burns, 91, 170
  cortex, 78
  disc, 41, 165
  gold, 154
  metallic, 42
  micro, 37
  monopolar, 51
  mouth, 93
  needle, 37
  pacemaker, 107
  percutaneous, 48
  pipette, 93, 126
  placements, 21, 22, 25, 51, 170,
       171, 175
  polarising, 93
  potential measuring, 24
  scalp, 51
  size, 22

skin, 87, 90
subcutaneous, 170, 174, 175
surface, 21, 40
transvascular, 108
types, 79

endocrine charges, 68

engram, 112

entropies, 5
epileptiform discharge, 93, 97

ethanol, 117

evoked potentials, 54, 117,
        119, 125
  responses, 33, 35, 37, 93

excitability,
  nerve, 61, 62, 63, 64
  thresholds, 27

exercise, 100

exocrinal side effects, 171

eyeball, 138

fatigue, 47, 48, 102

fatigability, 68

field, DC, 93, 97
  electric, 24
  electrostatic, 153
  magnetic, 9
  negative, 94
  potential, 33, 36

fluid dielectric, 173

fornix, 151, 161
fovial stimulation, 35

free radicals, 156

frequency, 40, 42, 43, 46, 64, 79

  carrier, 30
  high, 29, 43
  low, 39
frog, 57

gallamine, 168
gastric-cannula, 117

hand function, 101

headache, 68

hearing, 85

heart, 28,

heart rate, 31, 133

hemiplegia, 99

heroin, 139

hormone, 156

5-hydroxytryptamine, 80
hyperhidrosis, 68

hypertension, 170

hypotension, 68

hypothalamus, 82

impedance, 21, 25, 40, 41
  cephalography, 21
  fluctuations, 22
  measurement of tissue, 39, 42
      "   of skin, 43
  scalp, 40

inhibition, 156

inhibitory system, 135

insomnia, 140

interictal discharges, 93

intimal damage, 106

intravascular thrombogensis,
                105

ions, 9, 95

ionic currents, 9
  movement, 78

ionisation, 153

iontophoresis, 90

kidney infarction, 107

light, white, colored, 35

local burns, 170

loudness, 87

magnetice fields, 9
  detector, 10

magneto-inductive currents,
                153

magnetometers, 10

Maxwell's equations, 4

medial lemniscus, 117

memory, 111, 112

mesencephalon, 80, 81

micropipette, 126

microwave radiation, 57, 58

modulation, 45
modulated, signal, 31
  sine wave, 145, 159

momenta, 7

monkey, macaque, 117
  squirrel, 125

mono-polar electrodes, 51, 52

monosynaptic, 135

mouth electrode, 93

muscle,100
  forearm, 91
  relaxant, 168
  skeletal, 45

music, 86

negative field, 94

neonates, 133

neostigmine, 79

nerve, conduction, 9
  sciatic , 10, 57, 58, 59, 60,
            61,62
  peripheral, 117

neuronal activity, 71

obsessive, 139

occipital scalp, 35

oersteds, 13

optic tract, 126

orbital gyrus, 135

ozone, 153

pain, 89

Pantopaque, 174

paradoxical sleep, 162

paralysed,
  limbs, 45
  extremities, 99

paramagnetic molecules,156

Parkinson - like state, 79,
                        81

peroneal nerve, 47

penicillin, 94

phase, response, 5
  angle, 41

physostigmine, 79

physical therapy, 100

pinch strength, 101

placements, electrode, 22

polarizing electrode, 93

polysynaptic, 135

potentials, 7
  action, 58
  cortical, 37
  evoked, 54, 117, 119,125
  field, 33, 36
  measuring electrodes, 24
  scalp, 37
  surface, 21

probe, bipolar, 15
  Deutsch's, 16

psychiatry, 137

psychotherapy, electro, 141

pulse, 138

pulse, width, 46

pulsed current, 154

putamen, 82

radar, 57, 58

radiation, continuous wave,
          60, 62, 66
  microwave, 57

radicals, free, 156

radiophonic hearing, 85, 86

recovery of function, 99

recruiting response, 71

rectangular current,
  anesthesia, 126

reflexes, 101

relaxation, 142,
  muscle, 133

reserpine, 79

resistance, 5

resistivity, brain, 24
  cerebrospinal fluid, 18
  specific, 15, 18

respiration, 133, 168

response, evoked, 33, 35
  phase, 5

response, recruiting, 71

reticular activating
  system, 71
  formation, 136

retina, 125

retinal, ganglion fibres,
                    129
  evoked potentials, 125

Riemann's geometry, 7
  tensor, 3, 4

scalp, 52
  potentials, 37

schizophrenia, 139, 140

sciatic nerve, 10, 57, 58,
                  61, 62

seizure, 78

sequential stimulation, 45,
                        48

serotonin, 81

shielding, 10

shiver, 81

shock, convulsive, 27

signal modulated, 31

sine wave, modulated, 145,
                  148, 159

sinusoidal current, 16, 29,
                        30

skeletal muscle, 45

skin impedance, 43

skull, 52

sleep, 133, 142, 148, 154, 161
  depth, 155
  duration, 155
  electro, 135
  paradoxical, 162
  problems, 138

sleepiness, 68

solenoid, 154

somatosensory evoked potentials, 119
  system, 125

somnolence, 120

spasms, 170
  muscular, 99, 111

spatio-temporal map, 35, 37

specific resistivity, 15, 18

speech, 86, 87

spikes, 74

 splenic infarction, 107

stimulation, sequential, 45

stimulus fovial, 35
  frequency, high, 43
  frequency, low, 161
  thresholds, 29
  visual, 33

stomach, 117

strength, pinch, 101

stroke, 99

strychnine, 71, 74

substantia nigra, 81, 82

surface potentials, 21

synapse, first, 123

synaptic density, 125
  voltages, 95

systems, biological, 3

tail, clamp, 107

temporal, spatio, map, 35

tensor, Riemann's, 3,4,5

thalamus, 71, **78,** 82 , 123,
                  151, 161, 171,
                  173, 175

thalamic signals, 52

therapy, behavioral, 111
  electro, 15

thiopentone, 168

threshold of excitability,
                         64

thrombogenesis, 105

thrombus, 106

thyroid gland, 68

tibialis anticus muscle, 47

time, variability, 3

tissue, impedance, 39
  reaction, 48

tone, auditory, 57, 58
trains of current, 148, 159

transcranial stimulation, 37

transtemporal current, 154

tremor, 79

variability, time, 3

vascular changes, 68

vegetative changes, 68

vena cava, 106

ventricles, 18, 82, 133, 175

ventralis post. lat. nucleus,
         117, 123

vesicles, synapse, 125

vision, blurring, 143

visual, evoked response, 33
   patterns, 138, 142
   stimuli, 33
   system, 125

voltage, 3, 16
   measurement, 3
   synaptic, 96
   threshold, 31

vomiting, 170

white light, 35